Computational Algebraic Complexity

This publication has been reprinted from a special issue of the *Journal of Symbolic Computation*
Volume **9**, Number 3, March 1990

Computational Algebraic Complexity

Edited by

Erich Kaltofen
Department of Computer Science
Rensselaer Polytechnic Institute
Troy, USA

ACADEMIC PRESS
Harcourt Brace Jovanovich, Publishers
London San Diego New York
Boston Sydney Tokyo Toronto

ACADEMIC PRESS LIMITED
24/28 Oval Road,
London NW1 7DX

United States Edition published by
ACADEMIC PRESS INC.
San Diego, CA 92101

This book is printed on acid-free paper

British Library Cataloguing in Publication Data
is available

ISBN 0–12–394780–4

Printed and bound in Great Britain at the University Press, Cambridge

Contributors

E. Bach, Computer Sciences Department, University of Wisconsin-Madison, Madison, WI 53706, USA

J. Canny, 543 Evans Hall, Computer Science Division, University of California, Berkeley, CA 94720, USA

D. Coppersmith, Department of Mathematical Sciences, IBM Research Division, Thomas J. Watson Research Center, P.O. Box 218, Yorktown Heights, NY 10598, USA

E. Kaltofen, Department of Computer Science, Rensslaer Polytechnic Institute, Troy, NY 12180-3590, USA

H. Rolletschek, Johannes Kepler University, RISC Linz, A-4040 Linz, Austria

L. Ronyai, Computer and Automation Institute, Hungarian Academy of Sciences, P.O.B. 63, H-1502 Budapest, Hungary

V. Shoup, Computer Sciences Department, University of Wisconsin-Madison, Madison, WI 53706, USA

B.M. Trager, IBM, Thomas J. Watson Research Center, Yorktown Heights, NY 10598, USA

J. von zur Gathen, Department of Computer Science, University of Toronto, Toronto, Ontario M5S 1A4, Canada

S. Winograd, Department of Mathematical Sciences, IBM Research Division, Thomas J. Watson Research Center, Yorktown Heights, NY 10598, USA

R. Zippel, Department of Computer Science, Cornell University, Ithaca, NY 14853, USA

Preface

1. Introductory Remarks

The articles contained in this volume deal in a broad sense with the subject of complexity theory in algebraic computation. The idea to attract some of the best research contributions in this area of symbolic computation to a special volume on this subject was suggested to Erich Kaltofen by Bruno Buchberger after a meeting in December 1986. We then began the search and solicitation of papers that we thought would be suitable for such a volume. In terms of subject selection we were mostly guided by Strassen's (1984) survey of this research area. We do not hesitate in saying that the final selection of papers in this volume not only represents a surprising diversity of approaches and methods, but also contains—in our opinion—several of the most outstanding and significant contributions to the subject obtained in the past four years. Before we briefly discuss the individual contributions from our point of view, we would like to express our gratitude to all those who helped in the preparation of this volume, especially the authors who submitted their papers (despite an anticipated publication delay), and all referees, who must remain anonymous, for their constructive help in the selection process. We hope that with this volume the subject of computational algebraic complexity will remain a well represented topic in the *Journal of Symbolic Computation*.

2. Discussion of the Individual Contributions

The article by Bach and Shoup addresses a new question in the theory of randomized algorithms: is there a tradeoff between the number of random coin tosses needed and the probability of failure accomplishable, all while retaining polynomial expected running time? Bach and Shoup investigate the problem of factoring univariate polynomials over finite fields, which is one of the first problems in computer science on which the usefulness of randomization was demonstrated. They can show that the failure probability, that is the probability that their algorithm does not produce a full factorization of the input polynomial, can be kept exponentially small in the number of random bits used. As a consequence of this they can also obtain a more efficient deterministic method by just trying all the combinations for the few random bits requested. In particular, they obtained a deterministic algorithm that runs in polynomial time as a function of the input size times the squareroot of the characteristic of the coefficient field.

The article by Canny introduces a new invariant to the theory of affine polynomial ideals over a field, the generalized characteristic polynomial, abbreviated by GCP. While Cayley's last century multi-polynomial resultant captures the solvability of a zero-dimensional projective system of polynomial equations, no such invariant can exist for affine systems. However, projectivization of an affine system brings with it the possibility that a solution at infinity is no more zero-dimensional. Under this condition, the u-resultant of the projective solution, for instance, collapses to zero, obliterating all information on the isolated affine solution points. Canny introduces the GCP, which is, following Macaulay, the quotient of two characteristic polynomials rather than the quotient of the two determinants that constitute the resultant. He shows that the trailing coefficient of the GCP enjoys similar properties as the resultant, e.g., in the u-resultant case it contains as its factors all isolated affine zeros.

The article by Coppersmith and Winograd constitutes, as the first paper committed, the crystallization point of this special issue. The paper is on asymptotically fast matrix multiplication, starting from two previous results. One is an idea by Strassen (1988) that allows the coercion of a trilinear aggregate that does not correspond to a matrix multiplication scheme to one that does. The other is a theorem by Salem and Spencer on the density of certain subsets of initial blocks of the integers with the property that no three integers in the subset form an arithmetic progression. Coppersmith and Winograd implicitly construct a scheme for matrix multiplication that at the time of writing this preface still constitutes the frontier of the art of conceiving asymptotically fast methods. The reader will not only easily find the record breaking exponent of their algorithm in their paper, but also an open problem about the existence of a family of Abelian groups and subsets, which is seemingly unrelated to the matrix multiplication question, but whose resolution would imply the ultimate exponent 2.

The article by von zur Gathen improves on the functional polynomial decomposition algorithm published previously in the *Journal of Symbolic Computation* by Kozen and Landau (1989). Von zur Gathen's univariate solution is of sequential running time, as measured in coefficient field operations, a little bit above linear in the degree of the polynomial to be decomposed. Further results on a parallel version of the algorithm, a special bivariate factorization algorithm for so-called separated bivariate polynomials, and multivariate decomposition with the outer polynomial being univariate, complete the paper. This of von zur Gathen's work on the subject restricts itself to the so-called tame case, in which the degree of the outer function does not divide the characteristic of the coefficient field.

The article by the guest editor and Trager* focuses on a new representation of multivariate polynomials. The paper proposes to manipulate polynomials that are given by so-called black boxes, which are objects taking as input a value for each variable, and then producing the value of the polynomial they represent at the specified point. We show that within this representation the greatest common divisor and factorization problems are solvable in randomized polynomial time in the usual parameters, at least for coefficients from an algebraic extension of the rationals. The output of our algorithms are black boxes for the goal polynomials, for instance, programs for all the irreducible normalized factors. These outputs can be easily converted to the standard sparse format using modern sparse interpolation techniques (see the paper by Zippel this volume). They are also very compact and can be stored or transmitted easily. Lastly, we solve the somewhat more difficult problem of producing black box representations for the numerator and denominator of a multivariate rational function given by a black box.

The article by Rolletschek explores the efficiency of the Euclidean algorithm for those imaginary quadratic number rings that allow division with remainder. In an earlier paper published in the *Journal of Symbolic Computation* Rolletschek (1986) showed for the Gaussian integers that the most efficient way to obtain the greatest common divisor via a Euclidean remainder scheme, measured in terms of the number of division steps, is to use any of the norm-smallest remainder in every division. Here this optimality principle for the steepest descent scheme is extended to the discriminants -7 and -8 (the case of discriminant -3 having been settled previously by Lazard), while for -11 the principle is shown to be false: in fact, pairs of integers can be constructed in this number ring such that the steepest descent scheme requires arbitrarily more divisions than the optimal scheme does. Rolletschek introduces the use of finite state automata in the analysis of what regions of the complex plane the remainders of a standardized scheme can cycle through.

The article by Rónyai studies, from an algorithmic point of view, the Wedderburn–Artin structure theory of finite dimensional associative algebras over a finite field. The basic objects in this theory, such as the radical of such algebras, the decomposition of a semi-simple algebra into a direct product of simple ones, zero divisors, and the isomorphism from a simple algebra to a full matrix algebra, are all found in randomized polynomial time. Randomization is needed in all instances for solving arising polynomial factorization problems (see also the article by Bach and Shoup this volume), that is, the algorithms constitute deterministic polynomial time reductions to that problem. An application of the results to a problem on permutation groups is also exhibited.

* We wish to remark that the editorial process for this paper was handled entirely by Bruno Buchberger.

The article by Zippel deals with the problem of interpolating a sparse multivariate polynomial over fields of characteristic zero from its values. In his 1979 MIT Ph.D. thesis, Zippel originally conceived a method that used randomization and required, roughly speaking, cubic running time in the number of non-zero monomials present in the polynomial. Note that standard Lagrangian interpolation is exponential in the number of variables, independently on how many monomials there actually are in the polynomial. As it turns out, the special evaluations of Grigoryev and Karpinski (1987), used by Ben-Or and Tiwari (1988) in their sparse interpolation algorithm, are applicable to Zippel's approach as well, and result in a collection of surprisingly effective algorithms for sparse multivariate polynomial interpolation, which are presented here. Again estimating the efficiency coarsely, the best algorithms are now a bit above linear in the number of non-zero monomials present.

3. Present and Future Work

Clearly, this volume cannot encompass all areas of vigorous and fruitful investigation on the subject of algebraic computational complexity. For instance, the areas of the Gröbner bases theory or the theory of real elementary geometry have been covered by other special issues of the *Journal of Symbolic Computation* (vol. 5, nrs. 1&2, vol.6, nrs. 2&3), and lately complexity theoretic progress in these areas can be discerned (see, e.g., (Faugère et al. 1989) and (Renegar 1989)). We wish to point to some new work that advance several of the presented areas, and which we are aware of now. Shoup (1990) has been able to show that primitive elements in large finite fields can be selected using few random bits. Lakshman (1989) has made use of Canny's GCP to obtain complexity bounds for finding Gröbner bases for the radical of zero-dimensional affine ideals. Eberly (1989) in his recent Toronto Ph.D. thesis has developed the counterpart of Rónyai's theory to finite dimensional algebras over algebraic number fields. The sparse interpolation problem for arbitrary sample points has been investigated by Borodin and Tiwari (1989). Finally, the sparse interpolation methods in Zippel's article have been implemented by Wiley at Rensselaer Polytechnic Institute and demonstrated to be practical on polynomials with 100 monomials and more.

E. Kaltofen
B. Buchberger

4. Literature Cited

Ben-Or, M. and Tiwari P., "A deterministic algorithm for sparse multivariate polynomial interpolation," *Proc. 20th Annual ACM Symp. Theory Comp.*, pp. 301–309 (1988).

Borodin, A. and Tiwari, P., "On the decidability of sparse univariate polynomial interpolation," *Tech. Report* **RC 14923 (#66763)**, IBM Research Division, September 1989.

Eberly, W., "Computations for Algebras and Group Representations," *Tech. Report* **225**, Dept. Comput. Sci., Univ. Toronto, 1989.

Faugère, J.C., Gianni, P., Lazard, D., and Mora, T., "Efficient computation of zero-dimensional Gröbner bases by change of ordering," *Tech. Rep.* **89–52**, Univ. Paris 7, LITP, July 1989.

Grigoryev, D. Yu. and Karpinski, M., "The matching problem for bipartite graphs with polynomially bounded permanents is in NC," *Proc. 28th IEEE Symp. Foundations Comp. Sci.*, pp. 166–172 (1987).

Kozen, D. and Landau, S., "Polynomial decomposition algorithms," *J. Symbolic Comp.* **7/5**, pp. 445–456 (1989).

Lakshman, Y.N., "On the complexity of computing the Gröbner basis for the radical of a zero-dimensional ideal," *Proc. 22nd Annual ACM Symp. Theory Comp.*, to appear (1990).

Contents

Factoring Polynomials Using Fewer Random Bits

ERIC BACH AND VICTOR SHOUP†

Computer Sciences Department,
University of Wisconsin–Madison, Madison, WI 53706, U.S.A.

(*Received* 22 *May* 1989)

Let F be a field of $q = p^n$ elements, where p is prime. We present two new probabilistic algorithms for factoring polynomials in $F[X]$ that make particularly efficient use of random bits. They are easy to implement, and require no randomness beyond an initial seed whose length is proportional to the input size. The first algorithm is based on a procedure of Berlekamp; on input f in $F[X]$ of degree d, it uses $d \log_2 p$ random bits and produces in polynomial time a complete factorisation of f with a failure probability of no more than $1/p^{(1-\varepsilon)d/2}$. (Here ε denotes a fixed parameter between 0 and 1 that can be chosen by the implementer.) The second algorithm is based on a method of Cantor and Zassenhaus; it uses $d \log_2 q$ random bits and fails to find a complete factorisation with probability no more than $1/q^{(1-\varepsilon)d/4}$. For both of these algorithms, the failure probability is exponentially small in the number of random bits used.

1. Introduction

Let F be a finite field with q elements, where $q = p^n$ and p is prime. Consider the problem of factoring polynomials $f \in F[X]$ into irreducible factors. There are no known deterministic polynomial-time procedures for this problem, though there are efficient methods that use random numbers. We will present two algorithms that are easy to implement and make particularly efficient use of these random numbers. In particular, they require no randomness beyond an initial seed about as long as the input, and the probability of their not obtaining complete factorisations is exponentially small.

Berlekamp's factoring algorithm runs in time polynomial in $\deg f$ and q (Berlekamp, 1970). Thus, for bounded values of q, this is a polynomial-time algorithm. If q is arbitrary, however, Berlekamp's algorithm requires an exhaustive search through the field F and this takes exponential time. To avoid this, various *probabilistic* polynomial-time factoring algorithms have been invented (Berlekamp, 1970; Rabin, 1980; Ben-Or, 1981; Cantor & Zassenhaus, 1981). When endowed with the ability to flip coins, i.e. access to a source of independent, unbiased random bits, they are efficient and reliable.

In this paper we adopt the view that random bits are a scarce resource, and seek factoring algorithms that use this resource efficiently. There are two reasons for our point of view. First, a careful analysis of exactly where and how much randomness is required to solve a particular problem may produce valuable new insight into the problem. Second, whether polynomial-time computation benefits from randomness is an open question. One could conceivably resolve this in the negative by reducing the randomness requirement to zero. Our results may lead in this direction. To make our view precise, we introduce a formal definition of random bit usage that was proposed in Shoup (1987).

First, we adopt the following idealisation of probabilistic computation. On input x, an

† Victor Shoup's present address is AT&T Bell Laboratories, Murray Hill, NJ 07974, U.S.A.

algorithm A computes its random bit requirement $b(x)$ (the time to compute $b(x)$ and its value are polynomially bounded in $|x|$, the length of x). A then is supplied with a random bit string of length $b(x)$ and, after time bounded by a polynomial in $|x|$, it outputs an answer or the special symbol "?", signifying failure. We assume the failure probability $\varepsilon(x)$ is less than a fixed constant (1/2, say), and that if A succeeds in producing an answer, it is correct. Following Babai, we call such an algorithm a *Las Vegas* algorithm (see Johnson, 1984).

The measure we shall use to estimate random bit usage is the function $h(x)$ defined by

$$h(x) = \frac{b(x)}{\log_2 [1/\varepsilon(x)]}$$

if $\varepsilon(x)$ is nonzero and zero otherwise. $h(x)$ is called the *half-cost* of A on input x; it measures the relationship between the random bit consumption and the failure probability in an invariant fashion. In particular, it does not change if we iterate algorithm A to reduce the failure probability. Intuitively, the half-cost measures the number of random bits required to cut the failure probability in half, provided we buy random bits "in bulk". Since $b(x)$ is polynomial in $|x|$, so is $h(x)$. If $\varepsilon(x) > 0$, then $\varepsilon(x) \geq 2^{-b(x)}$, so $h(x) \geq 1$. Therefore, the best non-zero half-cost we could hope for is $h(x) = O(1)$. If $h(x) \leq H$ for all x, then $\varepsilon(x) \leq 2^{-b(x)/H}$, i.e. the failure probability is exponentially small in the number of random bits used.

Previously, Bach (1987) described a constant half-cost Las Vegas algorithm for finding square roots modulo primes. In this paper we extend this result and give two constant half-cost Las Vegas algorithms for polynomial factoring. The first is based on Berlekamp's algorithm and an extension described in Cantor & Zassenhaus (1981). Given a polynomial f of degree d in $F[X]$, it produces a *complete* factorisation of f using $d \log_2 p$ random bits, obtaining a failure probability bound of $1/p^{(1-\varepsilon)d/2}$. This implies a half-cost bound of $2/(1-\varepsilon)$ (here ε, $0 < \varepsilon < 1$, controls the tradeoff between running time and failure rate). The second algorithm is based on the distinct degree factorisation method (a so-called "folk" method) and improvements to this method described in Cantor & Zassenhaus (1981). It uses $d \log_2 q$ random bits, obtaining a failure probability bound of $1/q^{(1-\varepsilon)d/4}$. This implies a half-cost bound of $4/(1-\varepsilon)$. We do not claim that the first algorithm is necessarily superior to the second in practice, since the half-cost bounds contain an arbitrary parameter ε and randomness may not be the only resource of interest.

Both of these algorithms are efficient and easy to implement. The usual approach searches for a "splitting polynomial" h for which $\gcd(F(h), f) \neq 1, f$ (F is an easily computable function) and tries many independently chosen values of h. Our approach is to choose a seed at random and then deterministically generate a simple sequence that with very high probability contains a splitting polynomial. In particular, no sophisticated pseudo-random number generator is required.

We describe these algorithms in sections 2 and 3, respectively. In section 4, we briefly discuss some implementation details.

2. Modifications of the Berlekamp Algorithm

Recall that F is a finite field with $q = p^n$ elements, where p is prime. We assume that we are given a concrete representation $F = \mathbb{Z}_p(\alpha)$, where α is a root of an irreducible polynomial over $\mathbb{Z}_p$ of degree n.

Let f be a polynomial of degree d in $F[X]$ that we wish to factor. The first step of Berlekamp's algorithm is to obtain a polynomial f^* that has the same irreducible factors as f, but with each factor occurring only once, i.e. f^* is squarefree. This is easily done [see, e.g., Knuth (1981), p. 421]. So without loss of generality we will assume that the polynomial f is squarefree to begin with. Let $f=f_1 \cdots f_r$ be the complete factorisation of f.

F contains $\mathbb{Z}_p$ as a subfield. Furthermore, F is an n-dimensional $\mathbb{Z}_p$-vector space with $1, \alpha, \ldots, \alpha^{n-1}$ forming a basis. The ring $F[X]/(f)$ is an F-vector space of dimension d with the residues mod f of $1, \ldots, X^{d-1}$ forming a basis. Therefore, $F[X]/(f)$ is an nd-dimensional $\mathbb{Z}_p$-vector space with the residues mod f of $\alpha^i X^j$ $(i=0, \ldots, n-1; j=0, \ldots, d-1)$ forming a basis. Throughout the rest of this section, we purposefully blur the distinction between a polynomial and its residue mod f.

Now, by the Chinese Remainder Theorem, we have an isomorphism

$$F[X]/(f) \cong F[X]/(f_1) \times \cdots \times F[X]/(f_r).$$

For simplicity, we identify $F[X]/(f)$ with its isomorphic image, and call the residues

$$(a_1, a_2, \ldots, a_r)$$

the *components* of f. Since f_i is irreducible, $F[X]/(f_i)$ is a field containing $\mathbb{Z}_p$, the fixed field of the Frobenius automorphism $x \mapsto x^p$ on $F[X]/(f_i)$. Thus, $R=\prod_{i=1}^r \mathbb{Z}_p$ is an r-dimensional $\mathbb{Z}_p$-vector space, and is the kernel of the $\mathbb{Z}_p$-linear map on $F[X]/(f)$ given by $x \mapsto x^p - x$. Since we have an explicit basis for $F[X]/(f)$ over $\mathbb{Z}_p$, we can construct the matrix of this linear map, and by Gaussian elimination find a basis $\omega^{(1)}, \ldots, \omega^{(r)}$ for R over $\mathbb{Z}_p$.

If p is small, we can quickly factor f in the following manner. First, we compute

$$h_{ij} = \omega^{(i)} + j \quad (i=1, \ldots, r; j=0, \ldots, p-1).$$

Then, we initialise a set $S=\{f\}$, and iteratively refine it as follows: for each h_{ij} we consider each $\hat{f}$ in S in turn, and compute $\gcd(h_{ij}, \hat{f})$; if this is a proper divisor v of $\hat{f}$, we replace $\hat{f}$ in S by v and $\hat{f}/v$. Any pair of factors f_s, f_t must be separated by this procedure, since there must be some basis element

$$\omega^{(i)} = (w_1, \ldots, w_r) \quad (w_i \in \mathbb{Z}_p),$$

for which $w_s \neq w_t$ (otherwise, all elements of R would agree in their s and t components, which is not the case). For this $\omega^{(i)}$,

$$\omega^{(i)} - w_s \equiv 0 \pmod{f_s},$$
$$\omega^{(i)} - w_s \not\equiv 0 \pmod{f_t}.$$

Thus, $\omega^{(i)} - w_s$ will serve to separate f_s and f_t.

For large p, we can use a probabilistic method described in Cantor & Zassenhaus (1981) that goes as follows. Choose a random element h of R (this is easy as we have a $\mathbb{Z}_p$-basis for R). In component notation

$$h = (a_1, \ldots, a_r) \quad (a_i \in \mathbb{Z}_p).$$

Then compute

$$u = h^{(p-1)/2} = (a_1^{(p-1)/2}, \ldots, a_r^{(p-1)/2})$$

(note p is odd) and try to split f with $u-1$ and u. Let χ be the quadratic character on $\mathbb{Z}_p$. The probability that neither $\gcd(u-1, f)$ nor $\gcd(u, f)$ split f is the probability that $\chi(a_1) = \cdots = \chi(a_r)$, since each component of u is 0, 1, or -1. This probability is asymptotic to $1/2^{r-1}$ as $p \to \infty$. Using this method, we can construct a Las Vegas algorithm to completely factor f, but it will not have a constant half-cost. Indeed, if $f=f_1 f_2$, the probability that u splits f is about 1/2, leading to a half-cost near $2 \log_2 p$.

We now give a modification of this algorithm that *does* have a constant half-cost for sufficiently large p.

ALGORITHM 2.1. Input: f. Output: The set of irreducible factors of f.

(1) Construct the basis $\omega^{(1)}, \ldots, \omega^{(r)}$ for R over $\mathbb{Z}_p$ as described in paragraph 4 of this section.
(2) If $r=1$ then output $\{f\}$ and quit; otherwise, let $k=\lceil \frac{1}{2}\log_2 p\rceil$ and $S=\{f\}$. Repeat steps (3a)–(3d) d times.
(3a) Choose a random $x \in \mathbb{Z}_p$.
(3b) Let $h_{ij} = x\omega^{(i)} + j \quad (i=1, \ldots, r; j=0, \ldots, k)$.
(3c) For each h_{ij}, compute $u_{ij} = h_{ij}^{(p-1)/2}$.
(3d) For each u_{ij}, consider each $\hat{f}$ in S, and compute $\gcd(u_{ij}, \hat{f})$ and $\gcd(u_{ij}-1, \hat{f})$. If either of these is a proper divisor v of $\hat{f}$, then replace $\hat{f}$ in S by v and $\hat{f}/v$.
(4) If $|S|=r$, then output S; otherwise, report failure.

We now analyse the failure probability and half-cost of this algorithm. For fixed s and t, $1 \le s < t \le r$, let P_{st} be the probability that one iteration of step (3) of Algorithm 2.1 fails to separate f_s and f_t. We state the principal result of this section.

THEOREM 2.2. *We have*

$$P_{st} \le \frac{\log_2 p}{p^{1/2}}.$$

From this a half-cost bound (Corollary 2.5, below) easily follows. Before proving this we will need two lemmas, the first of which is proved in Schmidt (1976).

LEMMA 2.3. *Let χ be a multiplicative character of order $d>1$ on a finite field K with q elements. Suppose that $g(X) \in K[X]$ has m distinct roots (in the algebraic closure of K) and is not a dth power, i.e. not of the type $g(X)=c(h(X))^d$, where $c \in K$, and $h(X) \in K[X]$. Then*

$$\left| \sum_{x \in K} \chi(g(x)) \right| \le (m-1)q^{1/2}.$$

LEMMA 2.4. *If β, γ are distinct, nonzero elements of F, then none of the polynomials*

$$\prod_{1 \le j \le k} (X - j\beta)^{e_j}(X - j\gamma)^{e_j} \quad (0 \le e_j \le 1, \text{ not all } e_j \text{ zero})$$

are squares.

PROOF. Since β and γ are distinct, if such a polynomial were a square, then for distinct $j_1, \ldots, j_w$ between 1 and k,

$$\begin{aligned} j_1\beta &= j_2\gamma, \\ j_2\beta &= j_3\gamma, \\ &\vdots \\ j_{w-1}\beta &= j_w\gamma, \\ j_w\beta &= j_1\gamma. \end{aligned}$$

But this implies that $\beta \sum j_v = \gamma \sum j_v$, so either $\beta = \gamma$ or $\sum j_v = 0$. The first is impossible, and the last cannot be true because

$$0 < \sum_{\nu=1}^{w} j_\nu \le \sum_{j=1}^{k} j = k(k+1)/2 < p.$$

The last inequality follows from the fact that $k = \lceil \frac{1}{2} \log_2 p \rceil$. □

Proof of Theorem 2.2. Without loss of generality assume that $s = 1$, $t = 2$. There must be some basis element $\omega^{(i)} = (a, b, \ldots)$ for which $a \ne b$. If either a or b are zero, then h_{i0} will separate f_1 and f_2, provided $x \ne 0$. Otherwise, if $\chi(a) \ne \chi(b)$, where χ is the quadratic character, $h_{i0} - 1$ will separate f_1 and f_2, again, provided $x \ne 0$. In either case, we have $P_{12} \le 1/p$, so we can assume that a and b are nonzero and that $\chi(ab) = 1$.

For any j, $1 \le j \le k$, we cannot have $0 = ax + j = bx + j$, and if either $ax + j$ or $bx + j$ are zero, then h_{ij} will separate f_1 and f_2. So if we fail to separate f_1 and f_2, there must exist nonzero $y_1, \ldots, y_k$ in $\mathbb{Z}_p$ such that

$$\begin{aligned} (ax+1)(bx+1) &= y_1^2, \\ (ax+2)(bx+2) &= y_2^2, \\ &\vdots \\ (ax+k)(bx+k) &= y_k^2. \end{aligned}$$

Since $ab = c^2$, we can set $z_i = y_i/c$ and rewrite this system of equations as

$$\begin{aligned} (x+1/a)(x+1/b) &= z_1^2, \\ (x+2/a)(x+2/b) &= z_2^2, \\ &\vdots \\ (x+k/a)(x+k/b) &= z_k^2. \end{aligned} \tag{$*$}$$

Let $N = |\{(x, z_1, \ldots, z_k) \in \mathbb{Z}_p^{k+1} : (x, z_1, \ldots, z_k) \text{ satisfies } (*)\}|$. We want to get a good upper bound on N. Now, for fixed c, the number of z satisfying $z^2 = c$ is $1 + \chi(c)$. Therefore,

$$\begin{aligned} N &= \sum_{x \in \mathbb{Z}_p} \prod_{j=1}^{k} (1 + \chi((x+j/a)(x+j/b))) \\ &= \sum_{0 \le e_1, \ldots, e_k \le 1} \sum_{x \in \mathbb{Z}_p} \chi\left(\prod_{j=1}^{k} (x+j/a)^{e_j} (x+j/b)^{e_j} \right). \end{aligned}$$

In this last expression, the term corresponding to $e_1 = \cdots = e_k = 0$ is p. For the other terms, we can get an upper bound on the magnitude of the character sums appearing therein using Lemmas 2.3 and 2.4. These lemmas imply

$$N \le p + p^{1/2} \sum_{l=1}^{k} \binom{k}{l} (2l-1) = p + p^{1/2}(2^k(k-1)+1).$$

We divide this by 2^k to obtain a bound on the number of x for which there exist nonzero $z_1, \ldots, z_k$ satisfying $(*)$, and then divide by p to obtain a bound on the probability P_{12}. This produces

$$P_{12} \le 1/2^k + (k-1)/p^{1/2} + 1/(2^k p^{1/2}).$$

The right-hand side of this inequality is asymptotic to $\frac{1}{2} \log_2 p/p^{1/2}$ as $p \to \infty$, and some calculations show that it is less than $\log_2 p/p^{1/2}$ for all $p \ge 3$. So the theorem is proved. □

COROLLARY 2.5. *The failure probability of Algorithm* 2.1 *is bounded by*

$$\left(\frac{3^{4/3}(\log_2 p)^2}{p}\right)^{d/2} < 1/p^{(1-\varepsilon)d/2}.$$

(*the inequality holds for any* $\varepsilon > 0$ *provided* p *is sufficiently large*).

PROOF. Since the iterations of step (3) are independent, the probability of not separating any fixed pair of factors is at most

$$\left(\frac{(\log_2 p)^2}{p}\right)^{d/2}.$$

Summing over all pairs, we have a failure probability bound of

$$\begin{aligned}\binom{r}{2}\left(\frac{(\log_2 p)^2}{p}\right)^{d/2} &\leq d^2\left(\frac{(\log_2 p)^2}{p}\right)^{d/2}\\ &= \left(\frac{d^{4/d}(\log_2 p)^2}{p}\right)^{d/2}\\ &\leq \left(\frac{3^{4/3}(\log_2 p)^2}{p}\right)^{d/2}.\end{aligned}$$

The last inequality follows from the fact that $d^{1/d}$ is maximised at $d=3$.

This gives the first bound. To get the second, note that for any $\varepsilon > 0$ and p sufficiently large, we have $3^{4/3}(\log_2 p)^2 < p^{\varepsilon}$. □

The number of random bits used by Algorithm 2.1 is essentially $d \log_2 p$, so Corollary 2.5 implies that the half-cost is no more than $2/(1-\varepsilon)$, since for small p, we can resort to the deterministic version of Berlekamp's algorithm. This bound is valid even using the approximate uniform distribution described in section 4.

We note an interesting consequence of Theorem 2.2, due to the second author and worked out more fully in Shoup (1988). Since $P_{st} \leq \log_2 p/\sqrt{p}$, if more than $\sqrt{p}\log_2 p$ different values of x are used in step (3a) of Algorithm 2.1, one of them must split f. Hence, the time required to deterministically factor a polynomial over a finite field of characteristic p is at most a polynomial in the length of the input, times $\sqrt{p}$.

3. Modifications of the Cantor/Zassenhaus Algorithm

Again, F is a finite field with $q = p^n$ elements, and we want to factor $f \in F[X]$ of degree d. As in section 2, we assume that f is squarefree. The first step of the Cantor/Zassenhaus algorithm is to perform "distinct degree factorisation", that is, obtain factors $f^{(1)}, \ldots, f^{(m)}$ of f such that: (1) each $f^{(i)}$ is the product of r_i distinct irreducible polynomials of degree e_i; and (2) each irreducible factor of f appears in some $f^{(i)}$. This can be done in polynomial time by using the fact that $X^{q^a} - X$ is the product of all monic irreducible polynomials whose degrees divide a [see Cantor & Zassenhaus (1981) for details].

In this section, our analysis will involve several rings and fields, and it will be important to be explicit about the algebraic structures and homomorphisms under discussion. In particular, for $g, h \in F[X]$, we will use the notation $[g]_h$ to denote the image of g in $F[X]/(h)$.

We first discuss factoring $f = f_1 \cdots f_r$ where the f_is are distinct irreducible polynomials of degree e. We consider the case where p is odd in detail, and then sketch the differences for the case where $p = 2$.

By the Chinese Remainder Theorem, we have an isomorphism

$$F[X]/(f) \cong F[X]/(f_1) \times \cdots \times F[X]/(f_r).$$

Let $Q = q^e$. Each $F[X]/(f_i)$ is a field with Q elements containing F as a subfield. Fix some arbitrary field E of size Q that contains F. Then there is an isomorphism σ_i of $F[X]/(f_i)$ onto E that fixes F. Thus, we have an isomorphism of $F[X]/(f)$ onto E^r that sends $[h]_f$ to $(\sigma_1([h]_{f_1}), \ldots, \sigma_r([h]_{f_r}))$. We identify $F[X]/(f)$ with its isomorphic image.

The Cantor/Zassenhaus algorithm first chooses a polynomial h at random of degree $< re$. In component notation this will be

$$[h]_f = (a_1, \ldots, a_r),$$

where a_i is a randomly selected element of E. We compute $u = h^{(Q-1)/2} \bmod f$. Then

$$[u]_f = (a_1^{(Q-1)/2}, \ldots, a_r^{(Q-1)/2}).$$

Let χ be the quadratic character on E. The probability that neither $\gcd(u-1, f)$ nor $\gcd(u, f)$ split f is the probability that $\chi(a_1) = \cdots = \chi(a_r)$, which is asymptotic to $1/2^{r-1}$. Using this approach, we can construct a Las Vegas algorithm for factoring f, but it will certainly not have a constant half-cost.

We now give a new algorithm for factoring a polynomial f of this restricted type. Along with f, it is given as input an "iteration parameter" λ. We assume a canonical enumeration $C_1, C_2, \ldots$ of polynomials over F of degree $< e$.

ALGORITHM 3.1. Input: f, λ. Output: the set of irreducible factors of f.

(1) Let $k = \lceil \log_2 Q \rceil$, and $S = \{f\}$.
(2) Repeat steps (3a)–(3c) λ times.
(3a) Generate a random polynomial h in $F[X]$ of degree $< 2e$.
(3b) Compute $u_i = (h + C_i)^{(Q-1)/2} \bmod f$ $(i = 1, \ldots, k)$.
(3c) For each u_i, consider each $\hat{f}$ in S, and compute $\gcd(u_i, \hat{f})$ and $\gcd(u_i - 1, \hat{f})$. If either of these is a proper divisor v of $\hat{f}$, then replace $\hat{f}$ in S by v and $\hat{f}/v$.
(4) If $|S| = r$, then output S; otherwise, report failure.

We want to get a bound on the probability that Algorithm 3.1 fails to completely factor f. For fixed s and t, $1 \le s < t \le r$, let P_{st} denote the probability that one iteration of step (3) fails to separate f_s and f_t.

THEOREM 3.2. *We have*

$$P_{st} \le \frac{(\log_2 Q)^2}{Q}.$$

PROOF. Without loss of generality assume that $s = 1$, $t = 2$. We have an isomorphism of $F[X]/(f_1 f_2)$ onto E^2 which sends $[h]_{f_1 f_2}$ to $(\sigma_1([h]_{f_1}), \sigma_2([h]_{f_2}))$. We identify $F[X]/(f_1 f_2)$ with its isomorphic image. As h ranges over all polynomials of degree $< 2e$, it ranges over a complete residue system mod $f_1 f_2$. So if h is chosen at random, then in component notation, $[h]_{f_1 f_2} = (x, y)$, where x and y are randomly chosen elements of E.

For $i=1,\ldots,k$, let $a_i=\sigma_1([C_i]_{f_1})$ and $b_i=\sigma_2([C_i]_{f_2})$. The a_is are distinct, and so are the b_is. Now, $u_i\equiv(h+C_i)^{(Q-1)/2}\pmod f$, and so this congruence holds mod f_1f_2 as well. Therefore, in component notation

$$[u_i]_{f_1f_2}=((x+a_i)^{(Q-1)/2},(y+b_i)^{(Q-1)/2}).$$

Thus, P_{12} is no more than the probability that for $i=1,\ldots,k$, $\chi(x+a_i)=\chi(y+b_i)$, where we choose $x,y\in E$ at random. Let P'_{12} be the probability that for $i=1,\ldots,k$, $\chi(x+a_i)=\chi(y+b_i)\neq 0$. Then $P_{12}\leq k/Q^2+P'_{12}$. Now, P'_{12} is the probability that there exist nonzero $z_1,\ldots,z_k$ in E such that

$$\begin{aligned}(x+a_1)(y+b_1)&=z_1^2,\\(x+a_2)(y+b_2)&=z_2^2,\\&\vdots\\(x+a_k)(y+b_k)&=z_k^2.\end{aligned}\tag{$*$}$$

Let $N=|\{(x,y,z_1,\ldots,z_k)\in E^{k+2}:(x,y,z_1,\ldots,z_k)\text{ satisfies }(*)\}|$. We want to get a good upper bound on N. We have

$$\begin{aligned}N&=\sum_{x,y\in E}(1+\chi((x+a_1)(y+b_1)))\cdots(1+\chi((x+a_k)(y+b_k)))\\&=\sum_{0\leq i_1,\ldots,i_k\leq 1}\ \sum_{x,y\in E}\chi((x+a_1)^{i_1}(y+b_1)^{i_1}\cdots(x+a_k)^{i_k}(y+b_k)^{i_k})\\&=\sum_{0\leq i_1,\ldots,i_k\leq 1}\left(\sum_{x\in E}\chi((x+a_1)^{i_1}\cdots(x+a_k)^{i_k})\right)\left(\sum_{y\in E}\chi((y+b_1)^{i_1}\cdots(y+b_k)^{i_k})\right).\end{aligned}$$

In this last expression, the term corresponding to $i_1=\cdots=i_k=0$ is Q^2. We can use Lemma 2.3 to bound the magnitude of each of the other terms, obtaining

$$N\leq Q^2+Q\sum_{l=1}^{k}\binom{k}{l}(l-1)^2=Q^2+Q(k(k-1)2^{k-2}-k2^{k-1}+2^k-1).$$

We divide this quantity by 2^k to obtain a bound on the number of x, y for which there exist nonzero $z_1,\ldots,z_k$ satisfying ($*$), and by Q^2 to obtain a bound on P'_{12}. Using the fact that $P_{12}\leq k/Q^2+P'_{12}$, we have

$$P_{12}\leq\frac{1}{Q}\left(\frac{k}{Q}+\frac{Q}{2^k}+\frac{k(k-1)}{4}-\frac{k}{2}+1-\frac{1}{2^k}\right).$$

The right-hand side of this inequality is asymptotic to $\frac{1}{4}(\log_2 Q)^2/Q$ as $Q\to\infty$, and some calculations show that it is less than $(\log_2 Q)^2/Q$ for all $Q\geq 3$. So the theorem is proved. □

We now consider the case $q=2^n$. We will assume that F contains a primitive cube root of unity ω. If this is not the case, we can construct the quadratic extension of F using the polynomial $\omega^2+\omega+1$, factor f in this extension field, and then, if necessary, multiply conjugates together to obtain the factorisation of f over the original field.

We now modify Algorithm 3.1 to handle this case. Step (1) is changed so that

$k = \lceil \log_3 Q \rceil$. Step (3b) is changed so that

$$u_i = (h + C_i)^{(Q-1)/3} \bmod f.$$

In step (3c), we compute $\gcd(u_i, \hat{f})$, $\gcd(u_i + 1, \hat{f})$ and $\gcd(u_i + \omega, \hat{f})$.

We can prove an analog of Theorem 3.2 for this modified algorithm. Let $\hat{P}_{st}$ be the probability that one iteration of step (3) of this modified algorithm fails to separate f_s and f_t.

THEOREM 3.3. *We have*

$$\hat{P}_{st} \leq \frac{(\log_3 Q)^2}{Q}.$$

SKETCH OF PROOF. This is very similar to the proof of Theorem 3.2. The relevant system of equations is

$$(x + a_i)(y + b_i)^2 = z_i^3 \quad (i = 1, \ldots, k).$$

We can estimate the number of solutions by using the fact that for fixed $c \in E$, the number of $z \in E$ satisfying $z^3 = c$ is $1 + \chi(c) + \chi(c^2)$, where χ is a multiplicative character of order 3 on E. □

Using Algorithm 3.1, we now describe an algorithm for factoring any polynomial f in $F[X]$ of degree d. The basic idea is to perform distinct degree factorisation, and then apply Algorithm 3.1 to each of the resulting factors.

ALGORITHM 3.4. Input: f. Output: the set of irreducible factors of f.

(1) Perform distinct degree factorisation, obtaining $f^{(1)}, \ldots, f^{(m)}$, where $f^{(i)}$ has r_i factors, each of degree e_i. Initialise $S = \{\ \}$. Generate a list ρ of d random elements from F.
(2) For $i = 1, \ldots, m$, do the following. If $r_i = 1$, then add $f^{(i)}$ to S. Otherwise, let $\lambda_i = \lfloor d/(2e_i) \rfloor$. Run Algorithm 3.1 on $f^{(i)}$ with $\lambda = \lambda_i$, using ρ as the source of random field elements. If this succeeds in completely factoring $f^{(i)}$, then add the set of factors to S; otherwise, report failure.
(3) Output S.

To analyse this algorithm, we will ignore those $f^{(i)}$ with $r_i = 1$, since they are completely factored already. Therefore, to simplify the discussion, we will assume that $r_i \geq 2$ for each $i = 1, \ldots, m$. We will also assume that $e_1 < e_2 < \cdots < e_m$.

LEMMA 3.5. *Let $0 < \varepsilon < 1$ be a constant. Then there exists another constant C, depending on ε, such that for $i = 1, \ldots, m$, the probability that Algorithm 3.4 fails to completely factor $f^{(i)}$ is less than*

$$\frac{1}{q^{(1-\varepsilon)d/4}},$$

provided $q^{e_i} > C$.

PROOF. The probability that $f^{(i)}$ is not completely factored is no more than the sum over all s and t, $1 \leq s < t \leq r_i$, of P_{st}. By Theorems 3.2 and 3.3, this is at most

$$r_i^2 \left(\frac{(\log_2 Q_i)^2}{Q_i} \right)^{\lambda_i},$$

where $Q_i = q^{e_i}$ (note $\log_3 Q_i < \log_2 Q_i$). Since $\lambda_i \geq \lfloor r_i/2 \rfloor \geq (r_i - 1)/2$, we can bound this by

$$\left(\frac{r_i^{4/(r_i-1)}(\log_2 Q_i)^2}{Q_i}\right)^{\lambda_i} \leq \left(\frac{9(\log_2 Q_i)^2}{Q_i}\right)^{\lambda_i}.$$

Now, let ε, $0 < \varepsilon < 1$ be a constant. Then there exists a constant C such that $Q_i > C$ implies $9(\log_2 Q_i)^2 \leq Q_i^{\varepsilon}$. In this case, the failure probability is no more than $1/q^{(1-\varepsilon)e_i\lambda_i}$. We have $d = (2e_i)\lambda_i + \kappa_i$, where $0 \leq \kappa_i < 2e_i$. Since $d \geq 2e_i$, it follows that $(2e_i)\lambda_i > d/2$, and hence $e_i\lambda_i > d/4$. The lemma follows immediately. □

For fixed ε, we modify Algorithm 3.4 as follows. If $q^{e_i} \leq C$, then we factor $f^{(i)}$ deterministically by brute force examination of all monic polynomials of degree e_i, of which there are no more than C. Otherwise, we use Algorithm 3.1. The failure probability bound in Lemma 3.5 now holds unconditionally. Note that this bound holds even using the approximate distribution for ρ described in section 4.

THEOREM 3.6. *Let δ, $0 < \delta < 1$ be a constant. Then there exists another constant D, depending on δ, such that the failure probability of Algorithm 3.4 is less than*

$$\frac{1}{q^{(1-\delta)(1-\varepsilon)d/4}},$$

provided $q^{e_m} > D$.

PROOF. Summing over all $i = 1, \ldots, m$, we see that the probability of failing to factor some $f^{(i)}$ is at most $m/q^{(1-\varepsilon)d/4}$. For fixed $0 < \delta < 1$, we want to bound this quantity by $1/q^{(1-\delta)(1-\varepsilon)d/4}$. Suppose this bound fails. Then we have $q^d < m^\gamma$, where $\gamma = 4/(\delta(1-\varepsilon))$. We know that $d \geq 2e_1 + \cdots + 2e_m \geq m(m-1) + 2e_m$. So we have $q^{e_m} < (m^\gamma/q^{m(m-1)})^{1/2}$. Now, the quantity on the right-hand side goes to zero as $m \to \infty$, and so it is bounded by some constant D. □

We modify Algorithm 3.4 so that if $q^{e_m} \leq D$, then we factor f by brute force. The failure probability bound in Lemma 3.6 now holds unconditionally. The number of random bits used by Algorithm 3.4 is essentially $d \log_2 q$, implying a half-cost of no more than $4/((1-\delta)(1-\varepsilon))$.

4. Implementation Details

Throughout our discussion, we have tacitly assumed that by using about $e \log_2 p$ random bits, we can generate a random list of numbers

$$x = (x_1, \ldots, x_e) \quad (0 \leq x_i < p),$$

with a uniform distribution. We now justify this assumption.

We first point out that the "standard" method of generating a random number between 0 and p does not work. This method is to generate a random number using $\lceil \log_2 p \rceil$ random bits, and then throw this number away if it is too large. This process is repeated until a number in range is generated. This method uses far too many random bits for our purposes.

We now describe a method that does work. Let $b = \lceil \log_2 p^k \rceil$. Generate b random bits, and view the result as a number y, with $0 \leq y < 2^b$. Compute $z = y \bmod p^e$. Note that

$p^e \leq 2^b < 2p^e$. So for any $0 \leq z_0 < p^e$, the number of $0 \leq y_0 < 2^b$ such that $z_0 = y_0 \bmod p^e$ is at least 1 and at most 2. Therefore, for any z_0, $\Pr[z = z_0] \leq 2/2^b \leq 2/p^e$. Furthermore, $\Pr[z = z_0] \geq 1/2^b > 1/(2p^e)$.

Now, compute $x = (x_1, \ldots, x_e)$ where the x_is are the digits of z when z is written in base p. This gives a one-one correspondence between the possible values of z and the possible values of x. Therefore, for any e-tuple x_0, we have

$$\frac{1}{2}\left(\frac{1}{p^e}\right) < \Pr[x = x_0] \leq 2\left(\frac{1}{p^e}\right).$$

Suppose that an algorithm has a failure probability of α assuming a true uniform distribution. Let α' be the failure probability assuming our approximate uniform distribution. Let S be the set of e-tuples which cause failure. Then

$$\alpha' = \sum_{x_0 \in S} \Pr[x = x_0] \leq \sum_{x_0 \in S} 2/p^e = 2\alpha.$$

In all of our examples, α is sufficiently small that this increase by a factor of 2 is negligible.

This research was sponsored by the National Science Foundation, via grants DCR-8504485 and DCR-8552596.

References

Bach, E. (1987). Realistic analysis of some randomized algorithms. *Proc. 19th Annual ACM Symp. on Theory of Computing*, pp. 453–461. (Final version to appear, *J. Comput. Sys. Sci.*)

Ben-Or, M. (1981). Probabilistic algorithms in finite fields. *Proc. 22nd Annual Symp. on Foundations of Computer Science*, pp. 394–398.

Berlekamp, E. (1970). Factoring polynomials over large finite fields. *Math. Comp.* **24,** 713–735.

Cantor, D., Zassenhaus, H. (1981). A new algorithm for factoring polynomials over finite fields. *Math. Comp.* **36,** 587–592.

Johnson, D. (1984). The *NP*-completeness column: an ongoing guide. *J. Algorithms* **5,** 433–447.

Knuth, D. (1981). *The Art of Computer Programming, Vol. 2. Seminumerical Algorithms*, 2nd edn. Reading, MA: Addison-Wesley.

Rabin, M. (1980). Probabilistic algorithms in finite fields. *SIAM J. Comput.* **9,** 273–280.

Schmidt, W. (1976). *Equations over Finite Fields, Lecture Notes in Mathematics*, **536.** Berlin: Springer.

Shoup, V. (1987). *Finding Witnesses Using Fewer Random Bits*. University of Wisconsin–Madison, Computer Sciences Technical Report No. 725.

Shoup, V. (1988). *On the Deterministic Complexity of Factoring Polynomials over Finite Fields*. University of Wisconsin–Madison, Computer Sciences Technical Report No. 782; *Info. Process. Lett.*, to appear.

Generalised Characteristic Polynomials

JOHN CANNY*

543 Evans Hall, Computer Science Division, University of California, Berkeley

(*Received* 5 *October* 1987)

Multipolynomial resultants provide the most efficient methods known (in terms as asymptotic complexity) for solving certain systems of polynomial equations or eliminating variables (Bajaj *et al.*, 1988). The resultant of $f_1, \ldots, f_n$ in $K[x_1, \ldots, x_m]$ will be a polynomial in $m-n+1$ variables which is zero when the system $f_i = 0$ has a solution in $\bar{K}^m$ ($\bar{K}$ the algebraic closure of K). Thus the resultant defines a projection operator from $\bar{K}^m$ to $\bar{K}^{(m-n+1)}$. However, resultants are only exact conditions for homogeneous systems, and in the affine case just mentioned, the resultant may be zero even if the system has no affine solution. This is most serious when the solution set of the system of polynomials has "excess components" (components of dimension $>m-n$), which may not even be affine, since these cause the resultant to vanish identically. In this paper we describe a projection operator which is not identically zero, but which is guaranteed to vanish on all the proper (dimension $= m-n$) components of the system $f_i = 0$. Thus it fills the role of a general affine projection operator or variable elimination "black box" which can be used for arbitrary polynomial systems. The construction is based on a generalisation of the characteristic polynomial of a linear system to polynomial systems. As a corollary, we give a single-exponential time method for finding all the isolated solution points of a system of polynomials, even in the presence of infinitely many solutions at infinity or elsewhere.

1. Introduction

Our experience with linear algebra suggests that if we have n *polynomial* equations $f_i = 0$ in $K[x_1, \ldots, x_m]$, we should be able to successively eliminate variables and arrive at a single polynomial in $m-n+1$ variables which is zero when the original system has a solution. In fact successive elimination is an extremely inefficient way to arrive at such a condition, and it is possible to compute a much smaller polynomial whose vanishing is necessary and sufficient for the system $f_i = 0$ to have a solution. This polynomial is called the *resultant* or *eliminant* and can be thought of as a projection operator, since it defines an algebraic set which is the projection of the set defined by the system $f_i = 0$. This projection operator is a fundamental "black box" in algebraic computation with a variety of applications (Bajaj *et al.*, 1988). It can also be computed quite efficiently (Canny, 1987).

Unfortunately, although resultants are defined for homogeneous systems, they cannot readily be used for non-homogeneous systems. While the vanishing of the resultant is necessary and sufficient for a solution in projective space, in affine space we only get a necessary condition, i.e. the resultant as described in the last paragraph may vanish even if there is no affine solution, because of the presence of solution components "at infinity" (in a certain projective closure of the affine space). If these components are of excess ($>m-n$) dimension, the resultant will vanish identically, and provide no information about the

*Supported by a David and Lucile Packard Foundation Fellowship and by NSF Presidential Young Investigator Grant IRI-8958577.

solution set. In this paper we describe a projection operator that cannot vanish identically, and which is equal to the (homogeneous) resultant if the resultant is not identically zero.

The operator is based on a generalisation of the characteristic polynomial of a system of linear equations to systems of multivariate polynomial equations. The generalisation is natural in the sense that it reduces to the usual definition when all the polynomials are linear. Whereas the constant coefficient of the characteristic polynomial of a linear system is the determinant, the constant coefficient of the generalised characteristic polynomial (henceforth GCP) is the *resultant* of the system. More generally, the *lowest degree non-identically zero coefficient* of the GCP is the projection operator we are looking for. It has the property that it is zero on the projection of any *proper* (dimension $= m-n$) component of the algebraic set defined by the system $f_i = 0$.

The GCP projection operator has the following limitation: although it vanishes on the projection of all proper components, it may also vanish on the projection of certain proper algebraic sets that are contained in excess components. So if the lowest-degree coefficient of the GCP vanishes at a point, no information is given as to whether that point is the projection of a point in an excess or proper component. However, this information has not been needed in the applications of the GCP operator to date (Canny, 1987; Renegar, 1989; Ierardi, 1989).

As an application, we describe an efficient algorithm for solving systems of polynomial equations over the complex numbers which works even in the presence of infinitely many solutions "at infinity". The methods of Lazard (1981), Renegar (1987) and Canny (1987) all give single-exponential time bounds for the problem of solving polynomial systems. But these methods are all based on the u-resultant (van der Waerden, 1950) and are only applicable to systems to homogeneous polynomials having finitely many solutions. As we shall see in section 4 of this paper, the presence of an infinite number of solutions causes all the u-resultant based methods to fail. Using the GCP projection operator allows these methods to run in single-exponential time and to succeed even in the presence of an infinity of solutions. It does not matter whether these solutions are at infinity or elsewhere, and we obtain all the isolated points in the solution set.

When considering new equation solving and elimination problems, it is natural to contrast them with Gröbner basis algorithms (Buchberger, 1985). A lexicographic-ordered Gröbner basis can be very easily used to compute solutions of a polynomial system, and for variable elimination. However, it cannot be computed as efficiently as the GCP. The differences are most marked when the dimension of the set of solutions of the polynomial system is greater than one. Here the running time of the Gröbner basis algorithm may be double exponential (time $d^{2^{O(n)}}$) in the number of variables n, and there are examples of polynomials systems which attain this bound (Mayr & Meyer, 1982) (d is the maximum degree of the input polynomials). If the dimension of the set of solutions is zero, then much better bounds apply, and the basis algorithm will run in $d^{O(n^2)}$ time (Caniglia *et al.*, 1989). The example due to Brownawell (1987) suggests that this bound is tight. The method described in section 3 of this paper on the other hand, can find a point in every proper component of an algebraic set in time $d^{O(n)}$. If the input polynomials are densely represented, this bound is low order polynomial (about $O(N^4)$) in the input size for every fixed n. The Gröbner basis bound by contrast, even for dimension zero ($d^{O(n^2)}$), is super-polynomial for n fixed.

On sparse systems, Gröbner algorithms seem to do much better than the worst case bounds. For large, dense problems however, the resultant and GCP methods should be faster. Currently, the best resultant algorithm takes time $O(N^{2+\varepsilon})$ with dense input and fixed n (Canny *et al.*, 1989), and a good research topic is to reduce this to a pseudo-linear bound $O(N^{1+\varepsilon})$ for fixed n. Right now, there is a pseudo-linear algorithm only in the case $n = 2$. Resultant algorithms can be adapted to GCP calculation with an increase by a factor of N in the complexity. In fact GCP algorithms are not new. The methods of Renegar (1987) and Canny (1987) for resultant computation actually compute a GCP (or something very close to it) as a side effect. So it comes essentially free with these methods.

The paper is structured as follows. We give a definition of the generalised characteristic polynomial in section 2, and briefly sketch an algorithm for it. In section 3 we prove that the GCP has the desired properties, using some basic results about dimension of algebraic sets. We show that while the resultant may vanish identically in the presence of solutions of excess dimension, the lowest degree coefficient of the GCP still contains information about the components of proper dimension. Finally, in section 4 we apply these results to the equation solving problem. Using the u-resultant construction and the GCP we obtain a single exponential time algorithm which recovers all isolated solutions to a system of homogeneous polynomials even if the system has solutions of excess dimension.

2. Computation of Generalised Characteristic Polynomials

In this section we give the construction of the generalised characteristic polynomial $C(s)$ for a system of homogeneous polynomials f_i. It is a natural generalisation of the characteristic polynomial of a linear system and it equals the latter in the special case where all the f_i are linear. The constant coefficient of $C(s)$ is the (multipolynomial) resultant of the f_i. This property is analogous to the fact that the constant coefficient of the characteristic polynomial of a linear system is the determinant. Our construction is based on Macaulay's (1902) formula for the general resultant. Macaulay shows that the resultant equals the quotient of the determinant of a certain matrix A whose entries are coefficients of the polynomials, and a subdeterminant of A.

Suppose we are given n homogeneous polynomials f_i in n variables x_j, and that f_i has degree d_i. We need some notation for monomials of f_i. Let α be an n-tuple of integers, we write x^α for the monomial $x_1^{\alpha_1} \ldots x_n^{\alpha_n}$.

The rows and columns of the matrix A are indexed by the set of monomials in $x_1, \ldots, x_n$ of degree d where

$$d = 1 + \sum_{i=1,\ldots,n} (d_i - 1) \tag{1}$$

and letting X^d denote the set of monomials of degree d, the cardinality of X^d is

$$N = |X^d| = \binom{d+n-1}{d} \tag{2}$$

Definition. A polynomial is said to be *reduced* in x_i if its degree (the maximum degree of its monomials) in x_i is less than d_i. A polynomial that is reduced in all variables but one is said simply to be reduced.

Now consider the polynomial

$$F = C_1 f_1 + C_2 f_2 + \ldots + C_n f_n \tag{3}$$

where each C_i is a homogeneous polynomial of degree $d - d_i$ with symbolic coefficients, which is reduced in $x_1, \ldots, x_{i-1}$ for $2 \le i \le n$. F is a homogeneous polynomial of degree d, and so has N coefficients. There are also, in total, exactly N coefficients in the C_i. To see this, imagine for the moment that each f_i equals $x_i^{d_i}$. Then every monomial in F is a multiple of a monomial from exactly one of the C_is. For the monomial cx^α, let j be the smallest index i such that x^α is not reduced in x_i. Then cx^α is a multiple of a monomial from C_j and from no other C_i.

Since the coefficients of F are linear functions of the coefficients of the C_i via (3), this determines a linear map A from coefficients of the C_i to coefficients of F. Each non-zero entry in the matrix A is a coefficient of some f_i. This defines the matrix A that we mentioned earlier.

More concretely, if we index rows and columns of A by elements of X^d, then the column corresponding to x^α represents the polynomial

$$\frac{x^\alpha}{x_i^{d_i}} f_i \tag{4}$$

where i is the smallest j such that x^α has degree at least d_j in x_j.

The determinant of A vanishes if the f_i have a common zero, and it is therefore a multiple of the resultant R of the system (Macaulay, 1902). We can write $\det(A) = MR$, where M is an additional factor which we would like to remove. Macaulay shows that M is the determinant of a certain submatrix of A, in fact the submatrix of elements whose row and column indices are not reduced. Thus he obtains the simple formula $R = \det(A)/\det(M)$.

Having given a brief sketch of what a multipolynomial resultant is, we can now give the construction of generalised characteristic polynomials.

Definition. The *generalised characteristic polynomial* (or GCP), $C(s)$ of a system of homogeneous polynomials $f_1, \ldots, f_n$ in $x_1, \ldots, x_n$ is the resultant of $\hat{f}_1, \ldots, \hat{f}_n$, where $\hat{f}_i = f_i - sx_i^{d_i}$.

We do not claim this to be a novel construction. But what has not previously been observed is that it is both inexpensive to compute, and that it can be used to recover all the isolated zeros of a system of polynomials, as shown in the next section.

Inspection of the matrices A and M shows that the coefficients of $x_i^{d_i}$ in f_i always appear on the leading diagonals. So the determinant of the matrix $\hat{A}$ for the new system $\hat{f}_i$ is actually the characteristic polynomial (in the usual sense) of A, i.e. $\det(\hat{A}) = \det(A - sI) = \mathrm{CharPoly}(A)(s)$, where $\mathrm{CharPoly}(A)(s)$ denotes the characteristic polynomial of A in the variable s. The same holds true for M, so that the *generalised* characteristic polynomial of the f_i is given as

$$C(s) = \frac{\mathrm{CharPoly}(A)(s)}{\mathrm{CharPoly}(M)(s)} \tag{5}$$

Now A is an $N \times N$ matrix, while M has $N - D$ rows and columns, where

$$D = \sum_i \prod_{j \ne i} d_j \tag{6}$$

is the number of reduced rows (or columns). This follows because a reduced monomial is reduced in all variables but one, say the ith, and its degree in every other x_j is in the range $0, \ldots, d_j - 1$, i.e. d_j values. This implies that $\mathrm{CharPoly}(A)(s)$ has degree N and

CharPoly(M)(s) has degree $N-D$, so that the GCP $C(s)$ has degree D. To compute a characteristic polynomial using Newton's identity (Csanky, 1976) takes $O(N^4)$ arithmetic operations. For large problems, N is much larger than D, and so it seems that computation of all N coefficients of CharPoly(A)(s) in (5) is wasteful. But we can use the fact that if the quotient of two polynomials has degree D, then that quotient depends only on the D most significant coefficients of those polynomials. So it is possible to compute $C(s)$ by computing only the first D coefficients of CharPoly(A)(s) and CharPoly(M)(s). Using the Newton identity, this can be done with $O(N^3 D)$ operations.

3. Main Properties

We next prove our main result, that the GCP $C(s)$ contains all the information needed to recover the proper components of the zeros set of the f_i. This result gives as an immediate corollary, a method for finding all the zeros of a system of n non-homogeneous polynomials in n variables, even if such a system has infinitely many solutions "at infinity". The method is based on the u-resultant (van der Waerden, 1950), but unlike previous methods (Lazard, 1981; Reneger, 1987; Canny, 1987) does not require that there be only finitely many solutions at infinity. As noted previously, the GCP must be used with some care because it will also provide proper components within excess components, but does not distinguish these from the true proper components.

To begin, we give some definitions and basic results on dimension of algebraic sets. We will not define the dimension of an algebraic set, but detailed definitions are given in Mumford (1976) chapter 1, page 6. In what follows, we assume that variable values range over the complex numbers $\mathbb{C}$.

DEFINITION. The set of common zeros of a system of polynomials $f_1, \ldots, f_n$ in $x_1, \ldots, x_m$ is called an *algebraic set* and is denoted $V(f_1, \ldots, f_n) \subset \mathbb{C}^m$. An algebraic set $V(f)$ defined by a single polynomial (which is not identically zero) is called a *hypersurface*. If f is linear, then $V(f)$ is called a *hyperplane*.

If all the f_i are homogenous, it is more convenient to work with the projective space $\mathbb{P}^{m-1}$, formed by identifying points in $\mathbb{C}^m$ which are scalar multiples of each other. That is, a "point" in $\mathbb{P}^{m-1}$ corresponds to all points in $\mathbb{C}^m$ of the form $\lambda(p_1, \ldots, p_m)$, where the $p \in \mathbb{C}^m$ is a non-zero constant vector, and λ ranges over all non-zero complex values. Points in $\mathbb{P}^{m-1}$ are sometimes called solution "rays" for this reason. $\mathbb{P}^{m-1}$ has dimension $m-1$ and is compact. We use the same notation, $V(f_1, \ldots, f_n) \subset \mathbb{P}^{m-1}$ for an algebraic set defined by homogenous polynomials f_i.

DEFINITION. An algebraic set is said to be *reducible* if it can be expressed as a finite union of proper subsets which are algebraic. An algebraic set which is not reducible is *irreducible*.

Any algebraic set can always be expressed as a finite union of irreducible algebraic subsets called *components*. Many results in algebraic geometry apply only to irreducible algebraic sets, and in much of what follows, we work with the individual components of an algebraic set.

DEFINITION. Let Z be the intersection of m hypersurfaces in n-dimensional affine or projective space. A component W of Z is said to be *proper* if it has dimension $n-m$. A component of dimension greater than $n-m$ is said to be an *excess* component.

And in fact all components of an intersection must be either proper or excess by the following lemma:

LEMMA 3.1. *If f_i are m non-honogeneous polynomials in n variables, (or homogeneous in $n+1$ variables), then every component of $V(f_1, \ldots, f_m)$ has dimension at least $n-m$.*

For a proof, see for example Mumford (1976) corollary 3.14. Our main result is that if $C(s)$ is arranged in powers of s, then its lowest degree coefficient vanishes on the projection of all proper components of the intersection. We start with n polynomials $f_i(u_1, \ldots, u_m, x_1, \ldots, x_n)$, which are homogeneous in the x_j. Then:

THEOREM 3.2. *Let $Z = V(f_1, \ldots, f_n) \subset \mathbb{C}^m \times \mathbb{P}^{n-1}$, and let W be a proper component of Z, so that the dimension of W is $m-1$. Let $C(u_1, \ldots, u_m)(s)$ be the generalised characteristic polynomial of the f_i, as polynomials in the x_j. Arranging the GCP in powers of s, let $C_k(u_1, \ldots, u_m)$ be its coefficient of lowest degree. If $\pi_u : \mathbb{C}^m \times \mathbb{P}^{n-1} \to \mathbb{C}^m$ denotes projection on u_i-coordinates, then $C_k(\pi_u(p)) = 0$ for all $p \in W$.*

PROOF. The GCP is the resultant of the polynomials $\hat{f}_i = f_i - s x_i^{d_i}$. With the addition of the complex variable s, the zeros set of the $\hat{f}_i$, call it Z', lies in $\mathbb{C}^m \times \mathbb{P}^{n-1} \times \mathbb{C}$. Since $\hat{f}_i$ and f_i are identical when $s = 0$, the intersection of Z' and the hypersurface $s = 0$ is exactly $Z \times \{0\}$. So for every component W of Z, we have $W \times \{0\} \subset Z'$. If W is a *proper* component it has dimension $m-1$, but by the dimension lemma, every component of Z' has dimension at least m. So $W \times \{0\}$ must be contained in some component W' of dimension m.

Because every point of W' has an m-dimensional neighbourhood, and because the intersection of this neighbourhood with the hypersurface $s = 0$ is $(m-1)$-dimensional, it follows that for every point $p \in W \times \{0\}$, there is a sequence of points (p_j) in $W' - W \times \{0\}$ which converges to p. Writing $C(u_1, \ldots, u_m)(s)$ now for the GCP of the f_i, or equivalently the resultant of the $\hat{f}_i$, then $C(\pi_u(q))(\pi_s(q)) = 0$ for any point q in Z', where π_s denotes projection on the s-coordinate. In particular $C(\pi_u(p_j))(\pi_s(p_j)) = 0$ for all j. Dividing this polynomial through by $\pi_s(p_j)^k$ (which is non-zero), and letting C_i denote the coefficient of s^i in the GCP, we obtain

$$C_k(\pi_u(p_j)) + \sum_{i=k+1, \ldots, D} (\pi_s(p_j))^{i-k} C_i(\pi_u(p_j)) = 0 \tag{7}$$

for all p_j, where C_k is the lowest degree non-vanishing coefficient of $C(s)$, and D is the degree of $C(s)$. This expression is a polynomial in the coordinates of the p_j, and is therefore a continuous function of the coordinates. Since it is zero for all $p_j \to p$, it must be zero at p. But the point p has s-coordinate zero, so the summation over i vanishes, and we conclude that $C_k(\pi_u(p))$ must equal zero. □

We can restate the theorem succinctly as:

$$\pi_u(\cup W_i) \subset V(C_k) \subset \pi_u(V(f_1, \ldots, f_n)) \tag{8}$$

The second containment follows because $\pi_u(V(f_1, \ldots, f_n)) = V(C_0)$, where C_0 is the resultant of the f_i. If $k = 0$ it is trivially true, whereas $k > 0$ implies $C_0 = 0$, so that $V(C_0) = \mathbb{C}^m$.

CONJECTURE. We conjecture that if Z_i is *any* component of $V(f_1, \ldots, f_n)$, then it contains a component that intersects the closure of the perturbed system for $s \neq 0$, i.e. there exists a non-empty algebraic set U such that

$$Z_i \supset U \subset \overline{(V(\hat{f}_i, \ldots, \hat{f}_n) - V(s))} \tag{9}$$

To prove the conjecture, one needs to show that if Z is an excess component, for small enough ε, $V(\hat{f}_i, \ldots, \hat{f}_n) \cap V(s = \varepsilon)$ has a proper component "near" to Z. The intuition behind this is that if just one of the coefficients of $x_i^{d_i}$ in f_i is changed slightly, it causes each component of the intersection to either "move" slightly, or to be cut into components of lower dimension, which are all contained within that component. In either case, every point of the new intersection is close to some point of the old intersection. Applying this inductively to each f_i, we eventually obtain a new intersection with only proper components, such that each of its components is near to one of the original components.

It is also reasonable to conjecture that the degree of vanishing of $C(u_1, \ldots, u_m)(s)$, at some point $u_i = p_i$ is a measure of the intersection multiplicity (in some appropriate sense) of the surfaces defined by the f_i. For example, we could consider the intersection multiplicity of the surfaces $f_i(p_1, \ldots, p_m, x_1, \ldots, x_n) = v_i$ in $\mathbb{C}^{2n}$.

4. Application to Equation Solving

The main theorem of the last section can be applied to the following problem: Given n non-homogeneous polynomials g_i in n variables, $x_1, \ldots, x_n$ find all the isolated solution points of the system $g_i = 0$. By isolated solution points, we mean those points that are not contained in some higher-dimensional component of the solution set. The system has an equal number of equations and variables, and so the proper components of $V(g_1, \ldots, g_n)$ are zero-dimensional, i.e. points.

Since the methods we will use apply to homogeneous polynomials, we must produce a homogeneous system from the g_i by introducing an additional variable x_0. For each polynomial g_i of degree d_i we produce a homogeneous polynomial f_i of degree d_i by multiplying terms of g_i of degree δ_i by $x_0^{(d_i - \delta_i)}$. Then if $(p_1, \ldots, p_n) \in \mathbb{C}^n$ is a solution of the original system, $\lambda(1, p_1, \ldots, p_n) \in \mathbb{P}^n$ is a solution ray of the homogeneous system.

In fact there is a one-to-one correspondence between solution points of the original system and solution rays of the homogeneous system which have $x_0 \neq 0$. However, there may be solutions of the homogeneous system which have $x_0 = 0$, called "solutions at infinity" which have no counterpart in the original system. There may in fact be excess components of the intersection at infinity, even if the original system has only proper solutions.

The presence of excess components at infinity causes the methods of Lazard (1981), Renegar (1987) and Canny (1987) to fail, and there is no easy way to ensure that the given system has only proper solutions at infinity. The methods just mentioned are the only polynomial equation-solving methods that have single exponential bounds. They are based on the u-resultant which we now describe. Using the GCP, we can give a u-resultant style method with single-exponential time bounds which succeeds even in the presence of excess solutions at infinity or elsewhere.

To a system of n homogeneous polynomials f_i in $n+1$ variables, we add the linear polynomial

$$u_0 x_0 + u_1 x_1 + \ldots + u_n x_n \tag{10}$$

where the coefficients $u_0, \ldots, u_n$ are indeterminates. We call this last polynomial the *u-form* and denote it $U(x_0, \ldots, x_n, u_0, \ldots, u_n)$. We now have a system of $n+1$ homogeneous polynomials in $n+1$ variables, and the resultant of such a system is a polynomial $R(u_0, \ldots, u_n)$ called the u-resultant.

Suppose now that $\lambda(p_0, \ldots, p_n)$ is a solution ray of the system f_i. Then it will also satisfy the u-form U if and only if

$$p_0 u_0 + \ldots + p_n u_n = 0 \tag{11}$$

So the system as a whole has a solution, and therefore the resultant $R(u_0, \ldots, u_n)$ will vanish, whenever $p_0 u_0 + \ldots + p_n u_n = 0$. This implies that $(p_0 u_0 + \ldots + p_n u_n)$ *divides* $R(u_0, \ldots, u_n)$. Similarly, every other solution ray of the f_i leads to a corresponding linear factor of the u-resultant. By computing the u-resultant and factoring it over the complex numbers, we can obtain the coordinates of all the solution rays. This is the essence of the methods in Lazard (1981), Renegar (1987), and Canny (1987) although they differ in how the factorisation is computed.

But suppose now that $V(f_1, \ldots, f_n)$ has a component of dimension 1 (or higher). It is a standard result (Mumford, 1976, corollary 3.30), that two projective varieties in the same space always intersect if the sum of their dimensions is at least the dimension of the space. For any fixed set of values of the u_i, the equation $u_0 x_0 + \ldots + u_n x_n = 0$ defines a variety of dimension at least $n-1$ in $\mathbb{P}^n$, and this must always intersect an excess solution of the f_i, irrespective of the value of the u_i. So the polynomial $R(u_0, \ldots, u_n)$ must be zero for all values of the u_i, i.e. it is identically zero. This is why the u-resultant methods fail if there are excess components in the solution set.

To get around this problem, we use a slight variation of the GCP of the f_i and the u-form U. Normally in constructing the GCP we would subtract sx_n from the last polynomial, which is U. But x_n already has a symbolic coefficient (u_n) in U. We compute instead the resultant of $\hat{f}_i = f_i - sx_i^{d_i}$, for $i = 1, \ldots, n$ and the unchanged u-form, and this will still give us a non-vanishing resultant (since we could specialise $u_n = s$ which makes the resultant monic in s). Another way to compute this resultant is to compute the GCP of the f_i *and* U, and then to substitute $u_n + s$ for u_n in the GCP.

Let the resultant obtained by either method be denoted $R(s, u_0, \ldots, u_n)$, then we see that it is precisely the u-resultant of the system $\hat{f}_i = 0$. Now for almost all values of s, and in particular, all sufficiently small values of $s \neq 0$, the perturbed system $\hat{f}_i = 0$ will have a finite number of solution rays. This follows because the resultant is a non-identically vanishing polynomial in s, and is the u-resultant of the system $\hat{f}_i = 0$. The number of solutions will be $\prod d_i$ (counting multiplicity), by Bezout's theorem.

We can think of the zeros set $V(\hat{f}_i, \ldots, \hat{f}_n, U)$ as lying in the space $\mathbb{C}^{n+1} \times \mathbb{P}^n \times \mathbb{C}$ with coordinates $(u_0, \ldots, u_n, \lambda(x_0, \ldots, x_n), s)$. Taking a sequence $(s_k) \to 0$, each solution set $V(\hat{f}_i, \ldots, \hat{f}_n, U, s - s_k)$ consists of $\prod d_i$ n-dimensional planes, counting multiplicity, whose projections on u_i-coordinates are hyperplanes. Each of these hyperplanes must approach a limiting hyperplane (possibly via a subsequence) at $s = 0$. This follows because the Grassmanian of hyperplanes in $\mathbb{C}^{n+1}$ is compact (Mumford, 1976, p. 174).

If $C_k(u_0, \ldots, u_n)$ is the lowest degree non-vanishing coefficient of $R(s, u_0, \ldots, u_n)$ as a polynomial in s, then from the proof of the main theorem of the last section, C_k must vanish on the projection of all the limit points of $V(\hat{f}_i, \ldots, \hat{f}_n, U)$ as $s \to 0$. Thus it must vanish at all of the $\prod d_i$ limiting hyperplanes, and therefore factors completely into linear factors because the degree of C_k is also $\prod d_i$. To see this, recall that the degree of the resultant in the coefficients of each polynomial is the product of the degrees of the other polynomials. Since the u-form is independent of s, the degree of the resultant (and all its coefficients as a polynomial in s) in the u_is must be $\prod d_i$.

The equation solving methods of Renegar (1987) and Canny (1988) avoid explicit computation of $R(u_0, \ldots, u_n)$, since it has so many coefficients ($O(d^{n^2})$ if all polynomials have degree d). Instead, they compute certain specialisations of it. For example in Canny (1988) the solutions not at infinity can be found with the following specialisations: $R_0(v, t) = R(v, t, t^2, \ldots, t^n)$ and $R_i^+(v, t) = R(v, t, \ldots, t^i+1, \ldots, t^n)$ and $R_i^-(v, t) = R(v, t, \ldots, t^i-1, \ldots, t^n)$, for $i = 1, \ldots, n$. Making these specialisations *before* the resultant is computed means that all arithmetic is done on polynomials in two variables v and t, and so the number of coefficients is at most $O(d^{2n})$.

We can make the same specialisations of the u_is before making the perturbation. The arguments in Canny (1988) which show that the resultant is non-vanishing for the above specialisations also apply to the lowest degree coefficient of $R(s, u_0, \ldots, u_n)$. So it is impossible for example, that the lowest degree coefficient of the specialisation of $R(s, u_0, \ldots, u_n)$ could be some other coefficient than C_k. Since C_k factors completely into linear factors, the methods of Renegar (1987) and Canny (1988) for factorising the u-resultant from its specialisations still apply to C_k and its specialisations. So, to summarise, the isolated solution points of a system of polynomials can be found using Renegar (1987) or Canny (1988) by replacing each resultant with the lowest degree coefficient of a specialisation of $R(s, u_0, \ldots, u_n)$.

5. Conclusions

We described a new construction called the generalised characteristic polynomial, which is a useful adjunct to the multipolynomial resultant. The GCP can be used in situations where resultant-based methods fail because of the presence of components of excess dimension in the solution set of a system of polynomials. It provides a means for systematically perturbing a polynomial system away from a "bad" or excess intersection, and for recovering the proper components of the intersection, which are robust with respect to this perturbation. While it is guaranteed to contain projections of all proper components, it may also contain the projections of proper components that lie inside excess components.

The GCP can be obtained naturally from certain resultant algorithms. We showed that it can be computed as a quotient of the characteristic polynomials of two square matrices. By judicious use of Newton's identity for characteristic polynomials, the quotient can be found by computing only some of the coefficients of the matrix characteristic polynomials. This provides a significant reduction in the cost of computing the GCP. But there is still much that can be done to improve the running time of both resultant and GCP algorithms. Our algorithm required $O(N^4)$ operations as a function of the matrix size, whereas in the special case of two homogeneous polynomials, the (Sylvester) resultant can be computed with $O(N \log^2 N)$ operations. It should be possible to improve the GCP bounds to quadratic or pseudo-linear.

References

Bajaj, C., Garrity, T., Warren, J. (1988). On the applications of multi-equational resultants, Computer Science Department Technical Report No. 826, Purdue University.

Brownawell, D. (1987). Bounds for the degrees in the Nullstellensatz. *Ann. Math.* 2nd series, **126,** 577–591.

Buchberger, B. (1985). Gröbner: An algorithmic method in polynomial ideal theory. *Multidimensional Systems Theory* (N. K. Bose, ed.), D. Reidel, Dordrecht, chapter 6.

Caniglia, L., Galligo, A., Heintz, J. (1989). Some new effectivity bounds in computational geometry. *Proc. 6th Int. Conf. on Applied Algebra and Error-correcting codes,* LNCS 357, Springer-Verlag, Berlin, 131–152.

Canny, J. F. (1987). A new algebraic method for motion planning and real geometry. *Proc. 28th IEEE Symp.* FOCS, Los Angeles, 39–48.

Canny, J. F. (1988). Some algebraic and geometric computations in PSPACE. *Proc. ACM STOC,* Chicago, 460–467.

Canny, J., Kaltofen, E, Yagati, L. (1989). Solving systems of non-linear polynomial equations faster. *Proc. ACM Int. Symp. on Symbolic and Algebraic Computation,* Portland OR, 121–128.

Csanky, L. (1976). Fast parallel matrix inversion algorithms. *SIAM J. Comp.* **5,** 618–623.

Ierardi, D. (1989). Quantifier elimination in the theory of an algebraically-closed field. *Proc. 21st ACM STOC,* Seattle WA, 138–147.

Lazard, D. (1981). Résolution des systèmes d'équations algébriques. *Theor. Comp. Sci.* **15,** 146–156.

Macaulay, F. S. (1902). Some formulae in elimination. *Proc. London Math. Soc.* (1) **35,** 3–27.

Mayr, E., Meyer, A. (1982). The complexity of the word problem for commutative semigroups and polynomial ideals. *Adv. Math.* **46,** 305–329.

Mumford, D. (1976). *Algebraic Geometry I, Complex Projective Varieties.* Springer-Verlag, Berlin.

Renegar, J. (1987). On the worst case arithmetic complexity of approximating zeros of systems of polynomials. Technical Report, School of Operations Research and Industrial Engineering, Cornell University.

Renegar, J. (1989). On the computational complexity and geometry of the first-order theory of the reals, Parts I, II, and III, Technical Reports, School of Operations Research and Industrial Engineering, Cornell University.

van der Waerden, B. L. (1950). *Modern Algebra* (3rd edn.) F. Ungar, New York.

Matrix Multiplication via Arithmetic Progressions

DON COPPERSMITH and SHMUEL WINOGRAD

Department of Mathematical Sciences
IBM Research Division
Thomas J. Watson Research Center
P.O. Box 218
Yorktown Heights, New York 10598, U.S.A.

(*Received* 17 *May* 1987)

We present a new method for accelerating matrix multiplication asymptotically. This work builds on recent ideas of Volker Strassen, by using a basic trilinear form which is not a matrix product. We make novel use of the Salem-Spencer Theorem, which gives a fairly dense set of integers with no three-term arithmetic progression. Our resulting matrix exponent is 2.376.

1. Introduction.

A matrix multiplication algorithm is usually built as follows. First an algorithm for a small matrix problem is developed. Then a tensor product construction produces from it an algorithm for multiplying large matrices. Several times over the last two decades, the ground rules for constructing the basic algorithm have been relaxed, and with more care in the tensor product construction, it has been shown how to use these more lenient basic constructions to still give efficient algorithms for multiplying large matrices.

Recently Strassen (1986) found a new relaxation of the ground rules. His basic trilinear algorithm computes a trilinear form which is not a matrix product at all. In this trilinear form, the variables are collected into blocks. The block structure (the arrangement of the blocks) is that of a matrix product, and the fine structure (the arrangement of variables within individual blocks) is also that of a matrix product, but the overall structure is not, because the fine structures of different blocks are incompatible. After taking a tensor power of this trilinear form, Strassen operates on the block structure (that of a large matrix product) to reduce it to several block scalar multiplications. Each block scalar multiplication is itself a matrix product (the fine structure), so that he has several disjoint matrix products (sharing no variables). He can then apply Schönhage's τ-theorem to obtain an estimate of the matrix exponent ω:

$\omega < 2.479.$

Here we follow Strassen's lead. We use a basic trilinear algorithm closely related to Strassen's. The block structure of our trilinear form is not a matrix product, although the fine structure still is. We use a combinatorial theorem of Salem and Spencer (1942), which gives a fairly dense set of integers containing no three-term arithmetic progression. We hash the indices of the blocks of variables to integers, and set to zero any block of variables not mapping to the Salem-Spencer set. We do this in such a way that if the product $X^{[I]}Y^{[J]}Z^{[K]}$ is contained in our trilinear form, then the hash values $b_X(I)$, $b_Y(J)$, $b_Z(K)$ form an arithmetic progression. So for any product of nonzero blocks $X^{[I]}Y^{[J]}Z^{[K]}$ in our trilinear form, we will get $b_X(I) = b_Y(J) = b_Z(K)$. We choose parameters so that on average each nonzero block of variables is contained in at most one nonzero block product $X^{[I]}Y^{[J]}Z^{[K]}$, and set to zero some blocks of variables to ensure that this condition holds absolutely, not just on average. Then, as Strassen, we have several disjoint matrix products, and can apply Schönhage's τ-theorem to obtain our exponent

$\omega < 2.376.$

The rest of the paper is organized as follows. In Section 2 we review Schönhage's τ-theorem. In Section 3 we present Strassen's construction. Section 4 contains the results of the Salem-Spencer theorem. Section 5 presents an outline of the present construction. In Section 6 we present an example of our construction, which gives an exponent of 2.404. The version presented in Section 7 uses exactly the same ideas, but is complicated by more terms and more indices; this gives an improvement to of 2.388. Section 8 introduces yet more complicated techniques which achieve a slightly better estimate of 2.376. Section 9 contains some related ideas that were not as effective in reducing the exponent. We make miscellaneous remarks in Section 10.

Finally, Section 11 shows how the existence of a certain combinatorial construction would yield $\omega = 2$. We cannot tell whether this construction can be realized.

Earlier versions of this paper appeared as Coppersmith and Winograd (1986) and (1987).

Readers unfamiliar with previous work in matrix multiplication are referred to the excellent survey by Victor Pan (1984).

ACKNOWLEDGMENTS

We are grateful to James Shearer for the reference to Behrend's construction, which was the last constituent we needed for the present work. Victor Pan referred us to Salem and Spencer. Arnold Schönhage gave a more symmetric presentation of our starting algorithm in Section 6. James Davenport offered helpful comments on an early draft of the paper.

2. Schönhage's Theorem

The basic results from "classical" matrix multiplication can be summarized by Schönhage's τ-theorem:

Theorem (Schönhage): *Assume given a field F, coefficients $\alpha_{i,j,h,\ell}$, $\beta_{j,k,h,\ell}$, $\gamma_{k,i,h,\ell}$ in $F(\lambda)$ (the field of rational functions in a single indeterminate λ), and polynomials f_g over F, such that*

$$\sum_{\ell=1}^{L}\left(\sum_{i,j,h}\alpha_{i,j,h,\ell}x_{i,j}^{(h)}\right)\left(\sum_{j,k,h}\beta_{j,k,h,\ell}y_{j,k}^{(h)}\right)\left(\sum_{k,i,h}\gamma_{k,i,h,\ell}z_{k,i}^{(h)}\right)$$
$$=\sum_{h}\left(\sum_{i=1}^{m_h}\sum_{j=1}^{n_h}\sum_{k=1}^{p_h}x_{i,j}^{(h)}y_{j,k}^{(h)}z_{k,i}^{(h)}\right)+\sum_{g>0}\lambda^g f_g\left(x_{i,j}^{(h)},y_{j,k}^{(h)},z_{k,i}^{(h)}\right)$$

is an identity in $x_{i,j}^{(h)}, y_{j,k}^{(h)}, z_{k,i}^{(h)}, \lambda$. Then given $\varepsilon>0$, one can construct an algorithm to multiply $N\times N$ square matrices in $O(N^{3\tau+\varepsilon})$ operations, where τ satisfies

$$L=\sum_{h}(m_h n_h p_h)^{\tau}. \tag{1}$$

We will also write the error term as $O(\lambda)$, so that the hypothesis becomes

$$\sum_{\ell=1}^{L}\left(\sum_{i,j,h}\alpha_{i,j,h,\ell}x_{i,j}^{(h)}\right)\left(\sum_{j,k,h}\beta_{j,k,h,\ell}y_{j,k}^{(h)}\right)\left(\sum_{k,i,h}\gamma_{k,i,h,\ell}z_{k,i}^{(h)}\right)=\sum_{h}\left(\sum_{i,j,k}x_{i,j}^{(h)}y_{j,k}^{(h)}z_{k,i}^{(h)}\right)+O(\lambda).$$

Less formally, the hypothesis is a trilinear algorithm, using L bilinear multiplications to (approximately) compute simultaneously several independent matrix products, of dimension $m_h\times n_h$ times $n_h\times p_h$ (written $<m_h,n_h,p_h>$). The superscript (h) indicates which matrix the variable belongs to.

In such a situation, we define the **matrix exponent** obtained from the construction as $\omega=3\tau$.

Note: Here we have presented Schönhage's Theorem in its *trilinear* form, which will make our construction easier to describe. The *bilinear* version assumes L rank-1 bilinear forms

$$M_\ell \equiv \left(\sum_{i,j,h} \alpha_{i,j,h,\ell} x_{i,j}^{(h)}\right)\left(\sum_{j,k,h} \beta_{j,k,h,\ell} y_{j,k}^{(h)}\right), \quad \ell = 1, 2, \dots, L$$

connected by identities

$$v_{i,k}^{(h)} \equiv \sum_{j=1}^{n_h} x_{i,j}^{(h)} y_{j,k}^{(h)} = \sum_{\ell=1}^{L} \gamma_{k,i,h,\ell} M_\ell + O(\lambda), \quad i \le m_h, \ k \le p_h.$$

That is, one forms L bilinear products, each of which is a linear combination of x-variables times a linear combination of y-variables, and expresses the answers $v_{i,k}^{(h)}$ as linear combinations of these products, up to terms of order λ. The *answers* $v_{i,k}^{(h)}$ are viewed as duals to the variables $z_{k,i}^{(h)}$ in the trilinear presentation, and in fact the bilinear presentation is obtained from the trilinear one by identifying coefficients of $z_{k,i}^{(h)}$ in both sides of the equation.

3. Strassen's construction

Strassen has found a new relaxation of the ground rules for the construction of the basic algorithm, that is, he has relaxed the hypotheses of the theorem. A key element in his construction is the observation that, using the ability to multiply a pair of $N \times N$ matrices, one can "approximately" (in the λ sense) multiply $(3/4)N^2$ pairs of independent scalars, that is, compute

$$\sum_{i=1}^{(3/4)N^2} x_i y_i z_i + O(\lambda) \tag{2}$$

where all the x_i, y_i, z_i are independent. Namely, setting

$$g = [(3/2)(N+1)],$$

and multiplying each variable in

$$\sum_{i=1}^{N}\sum_{j=1}^{N}\sum_{k=1}^{N} x_{i,j} y_{j,k} z_{k,i}$$

by an appropriate power of λ, one obtains

$$\sum_{i=1}^{N}\sum_{j=1}^{N}\sum_{k=1}^{N} \left(x_{i,j}\lambda^{i^2+2ij}\right)\left(y_{j,k}\lambda^{j^2+2j(k-g)}\right)\left(z_{k,i}\lambda^{(k-g)^2+2(k-g)i}\right) = \sum_{\substack{i+j+k=g \\ 1 \le i,j,k \le N}} x_{i,j} y_{j,k} z_{k,i} + O(\lambda),$$

since the exponent of λ,

$$i^2 + 2ij + j^2 + 2j(k-g) + (k-g)^2 + 2(k-g)i = (i+j+k-g)^2,$$

is zero when $i+j+k=g$ and is positive otherwise. Since any two indices i,j uniquely determine the third $k=g-i-j$, each variable $x_{i,j}$ is involved in at most one product. There are about $\lceil (3/4)N^2 \rceil$ triples (i,j,k), $1 \le i,j,k \le N$, $i+j+k=g$. Call this the *matrix-to-scalar* construction.

Strassen uses the following basic trilinear identity, related to Victor Pan's "trilinear aggregation" (1978):

$$\sum_{i=1}^{q}\left(x_0^{[2]} + \lambda x_i^{[1]}\right)\left(y_0^{[1]} + \lambda y_i^{[2]}\right)\left(z_i\lambda^{-1}\right) + \left(x_0^{[2]}\right)\left(y_0^{[1]}\right)\left(-\sum_{i=1}^{q} z_i\lambda^{-1}\right) = \qquad (3)$$

$$= \sum_{i=1}^{q}\left(x_i^{[1]}y_0^{[1]}z_i + x_0^{[2]}y_i^{[2]}z_i\right) + O(\lambda).$$

This gives a basic algorithm, using $q+1$ multiplications to compute a block inner product:

$$\sum_{i=1}^{q}(x_i^{[1]}y_0^{[1]}z_i + x_0^{[2]}y_i^{[2]}z_i) + O(\lambda).$$

The superscripts denote indices in the block inner product, and are uniquely determined by the subscript indices. We can label x_i and x_0 with different superscripts because they are different variables; similarly y_i and y_0. But the z-variables are involved in both blocks. They are shared. This is the new complication in the basic algorithm. This algorithm does not in itself represent a matrix product.

(If we tried to represent this algorithm as a matrix product, we would find that since all the variables z_i are associated with the same y-variable $y_0^{[1]}$, they must all have the same k-index. But since they are all associated with the same x-variable $x_0^{[2]}$, they must all have the same i-index.)

The first block, $\sum_i x_i^{[1]}y_0^{[1]}z_i$, represents a matrix product of size $< q, 1, 1 >$. A $q \times 1$ matrix (column vector) x is multiplied by a 1×1 matrix (scalar) y_0 to yield a $q \times 1$ matrix (*column* vector) v, which is dual to the vector z. In the second block, $\sum_i x_0^{[2]}y_i^{[2]}z_i$ represents a matrix product of size $< 1, 1, q >$. A 1×1 matrix (scalar) x_0 is multiplied by a $1 \times q$ matrix (row vector) y to yield a $1 \times q$ matrix (*row* vector) v, which is again dual to the vector z. The difficulty comes when we try to add the two blocks. The indices i of v are "schizophrenic": they don't know whether to behave as row indices or as column indices. Strassen's construction gives a way out of this difficulty.

Take Construction (3) and the two constructions gotten by cyclic permutations of the variables x,y,z, and tensor them together, to get an algorithm requiring $(q+1)^3$ multiplications to compute

$$\sum_{i,j,k=1}^{q} \left(x_{i,j,0}^{[1,1]} y_{0,j,k}^{[1,1]} z_{i,0,k}^{[1,1]} + x_{i,j,k}^{[2,1]} y_{0,j,k}^{[1,1]} z_{i,0,0}^{[1,2]} + x_{i,j,0}^{[1,1]} y_{0,0,k}^{[1,2]} z_{i,j,k}^{[2,1]} + x_{i,j,k}^{[2,1]} y_{0,0,k}^{[1,2]} z_{i,j,0}^{[2,2]} +\right.$$
$$\left. + x_{0,j,0}^{[1,2]} y_{i,j,k}^{[2,1]} z_{i,0,k}^{[1,1]} + x_{0,j,k}^{[2,2]} y_{i,j,k}^{[2,1]} z_{i,0,0}^{[1,2]} + x_{0,j,0}^{[1,2]} y_{i,0,k}^{[2,2]} z_{i,j,k}^{[2,1]} + x_{0,j,k}^{[2,2]} y_{i,0,k}^{[2,2]} z_{i,j,0}^{[2,2]}\right) + O(\lambda).$$

This is a block 2×2 matrix product (indicated by the superscripts). Within each block is a smaller matrix product; for example the third block is the matrix product $\sum_{i,j,k=1}^{q} (x_{i,j,0}^{[1,1]} y_{0,0,k}^{[1,2]} z_{i,j,k}^{[2,1]})$, which can be interpreted as a matrix product of size $< q^2, 1, q >$:

$$\sum_{i,j,k=1}^{q} x_{(i,j),0} y_{0,k} z_{k,(i,j)},$$

with (i,j) acting as the I-index (shared by x and z and taking on q^2 values), 0 acting as the J-index, and k acting as the K-index.

Taking the Nth tensor power of this algorithm, one gets an algorithm, requiring $(q+1)^{3N}$ multiplications, and producing a block $2^N \times 2^N$ matrix product, each block of which is a matrix product of some size $<m,n,p>$ where $mnp = q^{3N}$. Applying the matrix-to-scalar construction to the block structure, one then obtains $(3/4)(2^N)^2$ independent matrix products, each of some size $<m,n,p>$ where $mnp = q^{3N}$. Applying the τ-theorem, one gets

$$\omega \leq 3\tau_N, \qquad (q+1)^{3N} = (3/4)2^{2N}(q^{3N})^{\tau_N}.$$

Taking N^{th} roots and letting N grow, the (3/4) becomes insignificant, and we have

$$\omega \leq 3\tau, \quad (q+1)^3 = 2^2 q^{3\tau}.$$

Letting $q = 5$, Strassen obtains

$$\omega \leq \log(6^3/2^2)/\log 5 = \log_5 54 \simeq 2.4785.$$

4. The Salem-Spencer Theorem

We will use the following theorem of Salem and Spencer (1942); see also Behrend (1946).

Theorem (Salem and Spencer): *Given* $\varepsilon > 0$, *there exists* $M_\varepsilon \simeq 2^{c/\varepsilon^2}$ *such that for all* $M > M_\varepsilon$, *there is a set* B *of* $M' > M^{1-\varepsilon}$ *distinct integers*

$$0 < b_1 < b_2 < \cdots < b_{M'} < M/2$$

with no three terms in an arithmetic progression:

$$\text{for } b_i, b_j, b_k \in B, \qquad b_i + b_j = 2b_k \quad \text{iff} \quad b_i = b_j = b_k.$$

We will be considering the ring $\mathbb{Z}_M$ of integers modulo M, where M is odd. Because the elements of the Salem-Spencer set satisfy $0 < b_i < M/2$, no three can form an arithmetic progression mod M:

$$\text{for } b_i, b_j, b_k \in B, \qquad b_i + b_j \equiv 2b_k \bmod M \quad \text{iff} \quad b_i = b_j = b_k. \tag{4}$$

5. New Construction: Outline

Previous authors in this field have exhibited their algorithms directly, but we will have to rely on hashing and counting arguments to show the existence of a suitable algorithm. We start with a modification of Strassen's starting algorithm, producing 3 (instead of 2) block products. We take the $3N^{th}$ tensor power, yielding $3^{3N} = 27^N$ block products. We will show that we can choose about $(27/4)^N$ out of these 27^N products, which are *independent* in the sense that a given block of variables $X^{[I]}$, $Y^{[J]}$, or $Z^{[K]}$ will occur in at most one of our chosen products, and *strongly independent* in the sense that for any block product $X^{[I]}Y^{[J]}Z^{[K]}$ in the original tensor power which is *not* in our chosen set, at least one of its blocks of variables ($X^{[I]}$, $Y^{[J]}$ or $Z^{[K]}$) will be absent from our chosen block products. Thus, by setting to zero any variables not in our chosen block products, we will set to zero all other products in the original tensor power. Once we have done that, we will use Schönhage's τ-theorem to provide an estimate of ω.

By indirect arguments we will show the existence of the strongly independent set of block products. We will start with a hash function from the set of block indices to the set of integers mod M ($\mathbb{Z}_M$), in such a way that if $X^{[I]}Y^{[J]}Z^{[K]}$ appears in the tensor power, the hashed indices $b_X(I)$, $b_Y(J)$, $b_Z(K)$ form a three-term arithmetic progression. We will use the Salem-Spencer theorem to control the existence of such arithmetic progressions. Finally, among the large class of hash functions available, a counting argument will show that at least one of them has the desired performance.

6. New Construction: Easy Case

Start with a modification of Strassen's basic algorithm (3); see also Pan (1978). We use $q+2$ multiplications:

$$\begin{aligned}
&\sum_{i=1}^{q} \lambda^{-2}\left(x_0^{[0]} + \lambda x_i^{[1]}\right)\left(y_0^{[0]} + \lambda y_i^{[1]}\right)\left(z_0^{[0]} + \lambda z_i^{[1]}\right) \\
&- \lambda^{-3}\left(x_0^{[0]} + \lambda^2 \sum x_i^{[1]}\right)\left(y_0^{[0]} + \lambda^2 \sum y_i^{[1]}\right)\left(z_0^{[0]} + \lambda^2 \sum z_i^{[1]}\right) \\
&+ \left(\lambda^{-3} - q\lambda^{-2}\right)\left(x_0^{[0]}\right)\left(y_0^{[0]}\right)\left(z_0^{[0]}\right) \\
&= \sum_{i=1}^{q}\left(x_0^{[0]} y_i^{[1]} z_i^{[1]} + x_i^{[1]} y_0^{[0]} z_i^{[1]} + x_i^{[1]} y_i^{[1]} z_0^{[0]}\right) + O(\lambda).
\end{aligned} \tag{5}$$

We have brought the factors λ^{-3}, $\left(\lambda^{-3} - q\lambda^{-2}\right)$ outside in order to reflect the symmetry.

Note. This is equivalent to the *bilinear* algorithm

$$\begin{aligned}
M_i &= \left(x_0^{[0]} + \lambda x_i^{[1]}\right)\left(y_0^{[0]} + \lambda y_i^{[1]}\right), \qquad i = 1, 2, \dots, q \\
M_{q+1} &= \left(x_0^{[0]} + \lambda^2 \sum x_i^{[1]}\right)\left(y_0^{[0]} + \lambda^2 \sum y_i^{[1]}\right) \\
M_{q+2} &= \left(x_0^{[0]}\right)\left(y_0^{[0]}\right) \\
v_i^{[1]} &\equiv x_0^{[0]} y_i^{[1]} + x_i^{[1]} y_0^{[0]} = \lambda^{-1} M_i - \lambda^{-1} M_{q+1} + O(\lambda), \qquad i = 1, 2, \dots, q \\
v_0^{[0]} &\equiv \sum_{i=1}^{q} x_i^{[1]} y_i^{[1]} = \sum_{i=1}^{q} \lambda^{-2} M_i - \lambda^{-3} M_{q+1} + \left(\lambda^{-3} - q\lambda^{-2}\right) M_{q+2} + O(\lambda)
\end{aligned}$$

where $v_j^{[J]}$ is the dual to the variable $z_j^{[J]}$, and the equivalence is gotten by identifying coefficients of $z_j^{[J]}$ in both sides of (5).

The x-variables break into two blocks: $X^{[0]} = \{x_0^{[0]}\}$ and $X^{[1]} = \{x_1^{[1]}, \dots, x_q^{[1]}\}$. Similarly the y-variables break into blocks $Y^{[0]}$ and $Y^{[1]}$, and the z-variables into blocks $Z^{[0]}$ and $Z^{[1]}$. When we *zero* a block $X^{[I]}$ (resp. $Y^{[J]}, Z^{[K]}$), we will set to zero all x - (resp. y-, z-) variables with the given index pattern.

Fix $\varepsilon > 0$. Select N large enough so that the M defined below will exceed M_ε from the Salem-Spencer Theorem.

Take the $3N^{th}$ tensor power of Construction (5). Each variable $x_i^{[I]}$ in the tensor power will be the tensor product of $3N$ variables $x_j^{[J]}$, one from each of $3N$ copies of the original algorithm. Its subscript i will be a vector of length $3N$ with entries in $\{0, 1, \dots, q\}$, made up of the $3N$ subscripts j. Its superscript $[I]$ will be a vector of length $3N$ with entries in $\{0, 1\}$, made up of the $3N$ superscripts $[J]$. As before, $[I]$ is uniquely determined by i.

Set to zero all variables $x_i^{[I]}$ except those for which I has exactly N indices of 0 and exactly $2N$ indices of 1; similarly y- and z-variables. Set $M = 2\binom{2N}{N} + 1$. Construct a Salem-Spencer set B. Select random integers $0 \le w_j < M, j = 0,1, \dots ,3N$. For each superscript $I \in \{0, 1\}^{3N}$, compute a *hash* as follows. For each of the $3N$ index positions j, let I_j denote the jth element of I (either 0 or 1). Define

$$b_X(I) \equiv \sum_{j=1}^{3N} I_j w_j (\bmod M)$$

$$b_Y(J) \equiv w_0 + \sum_{j=1}^{3N} J_j w_j (\bmod M)$$

$$b_Z(K) \equiv \left(w_0 + \sum_{j=1}^{3N} (2 - K_j) w_j \right) / 2 \ (\bmod M).$$

Since M is odd, division by 2 is well defined.

Notice that for any blocks $X^{[I]}, Y^{[J]}, Z^{[K]}$ whose product $X^{[I]}Y^{[J]}Z^{[K]}$ appears in the computed trilinear form, we have

$$b_X(I) + b_Y(J) - 2b_Z(K) \equiv 0 \mod M. \tag{6}$$

This follows by considering the contribution of each w_j, noticing that in the basic construction

$$I_j + J_j + K_j = 2$$

for each of the three terms $x_0^{[0]} y_i^{[1]} z_i^{[1]}, x_i^{[1]} y_0^{[0]} z_i^{[1]}, x_i^{[1]} y_i^{[1]} z_0^{[0]}$.

Set to zero all blocks $X^{[I]}$ for which $b_X(I)$ is not in B. Similarly set to zero all blocks $Y^{[J]}$ for which $b_Y(J)$ is not in B, and blocks $Z^{[K]}$ for which $b_Z(K)$ is not in B. Then for any nonzero term $X^{[I]}Y^{[J]}Z^{[K]}$ remaining in our construction, we have

$$b_X(I) + b_Y(J) \equiv 2b_Z(K) \bmod M, \qquad b_X(I), b_Y(J), b_Z(K) \in B,$$

so that

$$b_X(I) = b_Y(J) = b_Z(K),$$

by the properties of B.

For each element $b \in B$ in the Salem-Spencer set, make a list of triples $(X^{[I]}, Y^{[J]}, Z^{[K]})$ of compatible nonzero blocks, with $b_X(I) = b_Y(J) = b_Z(K) = b$. (A *block* $X^{[I]}$ is the set of q^{2N} variables $x_{I'}^{[I]}$ with nonzero indices in $2N$ specified places, that is, sharing a common superscript I. A *nonzero block* is one which has not yet been set to zero. Blocks $X^{[I]}$, $Y^{[J]}$, $Z^{[K]}$ are *compatible* if the locations of their zero indices are pairwise disjoint.) For each triple $(X^{[I]}, Y^{[J]}, Z^{[K]})$ on the list, if it shares a block (say $Z^{[K]}$) with another triple $(X^{[I']}, Y^{[J']}, Z^{[K]})$ occurring earlier in the list, we set to zero one of the other blocks (say $Y^{[J]}$), and thus eliminate this triple. (If each of $X^{[I]}$, $Y^{[J]}$, $Z^{[K]}$ is shared with previous triples, we will end up eliminating at least two triples by zeroing one block of variables.)

For a fixed element $b \in B$, the expected number of triples in the list, before pruning, is

$$\binom{3N}{N,N,N} M^{-2}.$$

Here $\binom{3N}{N,N,N}$ represents the number of compatible triples $(X^{[I]}, Y^{[J]}, Z^{[K]})$ and the M^{-2} represents the probability of the (independent) events $b_X(I) = b$ and $b_Y(J) = b$. (If both hold, then $b_Z(K) = b$ follows.) That is, for fixed blocks $X^{[I]}$, $Y^{[J]}$, and fixed integer $b \bmod M$, if we randomize the values $w_0, w_1, \ldots, w_{3N}$, then

$$\mathrm{Prob}\{b_X(I) = b_Y(J) = b\} = \mathrm{Prob}\{b_X(I) = b\}\mathrm{Prob}\{b_Y(J) = b\} = M^{-1}M^{-1} = M^{-2},$$

since the sums $b_X(I)$ and $b_Y(J)$ involve different random variables. The expected number of compatible triples $(X^{[I]}, Y^{[J]}, Z^{[K]})$ with $b_X(I) = b_Y(J) = b_Z(K) = b$ is the sum of these probabilities (M^{-2}) over the $\binom{3N}{N,N,N}$ possible triples. We do not need independence among triples, since the expected value of a sum of random variables is the sum of their expected values, regardless of independence.

The expected number of unordered pairs $(X^{[I]}, Y^{[J]}, Z^{[K]})$, $(X^{[I']}, Y^{[J']}, Z^{[K]})$ sharing a Z-block is

$$(1/2)\binom{3N}{N,N,N}\left(\binom{2N}{N,N} - 1\right)M^{-3}.$$

Again $\binom{3N}{N,N,N}$ counts the compatible triples $(X^{[I]}, Y^{[J]}, Z^{[K]})$. The binomial coefficient $\binom{2N}{N,N} - 1$ counts the blocks $Y^{[J']}$ compatible with $Z^{[K]}$ (other than $Y^{[J]}$ itself). The factor 1/2 eliminates duplicate entries $((X^{[I]}, Y^{[J]}, Z^{[K]}), (X^{[I']}, Y^{[J']}, Z^{[K]}))$ and $((X^{[I']}, Y^{[J']}, Z^{[K]}), (X^{[I]}, Y^{[J]}, Z^{[K]}))$. The factor M^{-3} is the probability of the independent events $b_Z(K) = b$, $b_Y(J) = b$, $b_Y(J') = b$. They are independent even if indices are equal $(J' = I)$, because of the presence of the random variable w_0.

The expected number of pairs of triples sharing $Y^{[J]}$, or sharing $X^{[I]}$, is the same.

Suppose we eliminate a block $(Y^{[J]})$ because of a pair of triples $((X^{[I]}, Y^{[J]}, Z^{[K]}), (X^{[I']}, Y^{[J']}, Z^{[K]}))$ sharing a Z-block. If L triples (not yet eliminated) share this Y-block, then setting $Y^{[J]}$ to zero eliminates these L triples, while eliminating at least $\binom{L}{2} + 1$ pairs, namely all those sharing $Y^{[J]}$, and at least the pair sharing $Z^{[K]}$. Since $\binom{L}{2} + 1 \geq L$, we eliminate at least as many pairs as triples. Thus:

Lemma. *The expected number of triples remaining on each list, after pruning, is at least*

$$\begin{aligned} &\binom{3N}{N,N,N} M^{-2} - (3/2)\binom{3N}{N,N,N}\left[\binom{2N}{N,N} - 1\right] M^{-3} \\ &\geq (1/4)\binom{3N}{N,N,N} M^{-2}. \end{aligned} \tag{7}$$

The expected number of triples remaining on all lists, after pruning, is at least

$$H \equiv (1/4)M'\binom{3N}{N,N,N} M^{-2}. \tag{8}$$

This expectation H is an average over the choices of w_j. There is a choice of w_j which achieves at least H; fix such a choice.

Our algorithm computes at least H block scalar products $X^{[I]} Y^{[J]} Z^{[K]}$. The fine structure of each block scalar product is in fact a matrix product of size

$$< q^N, q^N, q^N >,$$

and all the variables are disjoint (by the Salem-Spencer property). From the τ-theorem we obtain

$$\omega \leq 3\tau_N, \quad (q+2)^{3N} \geq (1/4)M'\binom{3N}{N,N,N} M^{-2} q^{3N\tau_N}.$$

Use Stirling's approximation to obtain

$$(q+2)^{3N} \geq cN^{-1/2+\varepsilon} 3^{3N} 2^{-2N(1+\varepsilon)} q^{3N\tau N},$$

where c is a constant. Letting ε go to zero and N to infinity, and taking N^{th} roots, we obtain

$$(q+2)^3 \geq (3^3/2^2) q^{3\tau}$$

$$\omega \leq 3\tau \leq \log_q\left(\frac{4(q+2)^3}{27}\right).$$

Setting $q = 8$ we obtain

$$\omega \leq \log_8(4000/27) < 2.40364. \tag{9}$$

7. New Construction: Complicated Version.

In this section and the sequel, we will improve the exponent to 2.388 and then 2.376, by using the same ideas as in the previous section, on more complicated starting algorithms.

Begin with the basic algorithm:

$$\begin{aligned}
&\sum_{i=1}^{q} \lambda^{-2}\left(x_0^{[0]} + \lambda x_i^{[1]}\right)\left(y_0^{[0]} + \lambda y_i^{[1]}\right)\left(z_0^{[0]} + \lambda z_i^{[1]}\right) \\
&- \lambda^{-3}\left(x_0^{[0]} + \lambda^2 \sum x_i^{[1]}\right)\left(y_0^{[0]} + \lambda^2 \sum y_i^{[1]}\right)\left(z_0^{[0]} + \lambda^2 \sum z_i^{[1]}\right) \\
&+ \left[\lambda^{-3} - q\lambda^{-2}\right]\left(x_0^{[0]} + \lambda^3 x_{q+1}^{[2]}\right)\left(y_0^{[0]} + \lambda^3 y_{q+1}^{[2]}\right)\left(z_0^{[0]} + \lambda^3 z_{q+1}^{[2]}\right) \\
&= \sum_{i=1}^{q}\left(x_0^{[0]} y_i^{[1]} z_i^{[1]} + x_i^{[1]} y_0^{[0]} z_i^{[1]} + x_i^{[1]} y_i^{[1]} z_0^{[0]}\right) + \\
&x_0^{[0]} y_0^{[0]} z_{q+1}^{[2]} + x_0^{[0]} y_{q+1}^{[2]} z_0^{[0]} + x_{q+1}^{[2]} y_0^{[0]} z_0^{[0]} + O(\lambda).
\end{aligned} \tag{10}$$

The subscripts now form three classes: $\{0\}$, $\{q+1\}$, and $\{1, 2, \ldots, q\}$, which will again be denoted i. Again the subscripts uniquely determine the superscripts (block indices).

Take the $3N^{th}$ power of this construction. Set $L = [\beta N]$ (greatest integer), where β will be determined later. Set to zero all blocks of variables except those whose superscripts contain exactly $N + L$ indices of 0, $2N - 2L$ indices of 1, and L indices of 2.

Set

$$M = 2\binom{N+L}{L, L, N-L}\binom{2N-2L}{N-L, N-L} + 1.$$

Let I_j pick out the jth index of I as before. Define $b_X(I)$, $b_Y(J)$, $b_Z(K)$ as before, and set to zero any block of variables with $b_X(I)$ (resp. $b_Y(J)$, $b_Z(K)$) not in the Salem-Spencer set. For each b in the Salem-Spencer set, make a list of triples $(X^{[I]}, Y^{[J]}, Z^{[K]})$ of blocks, with $b_X(I) = b_Y(J) = b_Z(K) = b$, and eliminate entries with duplicated blocks.

For a given block $Z^{[K]}$, the number of pairs of blocks $(X^{[I]}, Y^{[J]})$ compatible with $Z^{[K]}$ is

$$\binom{N+L}{L,L,N-L}\binom{2N-2L}{N-L,N-L},$$

since the $N+L$ indices of 0 in K correspond to L instances of (0 in I, 2 in J), L instances of (2 in I, 0 in J), and $N-L$ instances of (1 in I, 1 in J); the $2N-2L$ indices of 1 in K correspond to $N-L$ instances of (1 in I, 0 in J) and $N-L$ instances of (0 in I, 1 in J); and the L instances of 2 in K all have (0 in I, 0 in J). Since M is twice this size, the elimination of duplicates proceeds as before and leaves a constant fraction of the triples intact.

We have M' lists, each with (on average) at least

$$(1/4)\binom{3N}{L,L,L,N-L,N-L,N-L}M^{-2}$$

entries, all having independent variables. (The multinomial coefficient indicates that there are L instances of (2,0,0) as $(x-,\ y-,\ z-)$indices, L of (0,2,0), L of (0,0,2), $N-L$ of (1,1,0), etc.). Each entry corresponds to a matrix product of size

$$<q^{N-L}, q^{N-L}, q^{N-L}>.$$

Thus our equation is

$$(q+2)^{3N} \geq (1/4)M'\binom{3N}{L,L,L,N-L,N-L,N-L}M^{-2}q^{3(N-L)\tau_N}$$

$$\simeq cN^{(-1+3\varepsilon/2)}\left[\frac{27}{\beta^{\beta}(1+\beta)^{1+\beta}(2-2\beta)^{2-2\beta}}\right]^N q^{3N(1-\beta)\tau_N}c'^{\varepsilon N}.$$

Letting ε tend to zero and N to infinity, and taking N^{th} roots, we get

$$(q+2)^3 \geq \frac{27}{\beta^{\beta}(1+\beta)^{1+\beta}(2-2\beta)^{2-2\beta}}q^{3(1-\beta)\tau}.$$

For $q=6,\ \beta=0.048$, we find

$$\omega \leq 3\tau < 2.38719.$$

8. Coupling the Weights

So far we have assumed that the weights w_j are independent random variables. In this section we will make them dependent: essentially $w_{2j-1} = w_{2j}$. One consequence is that the randomness arguments need to be redone. The advantage is that we are able to gain higher estimates of the "value" of various pieces.

It will be convenient to have a notion of the "value" of a trilinear form (or trilinear algorithm). We define the "value" V_τ of a trilinear form A in terms of the matrix products it can simulate. Suppose the matrix exponent is $\omega = 3\tau$. Suppose that a tensor power of A can be reduced (by substitution of variables) to the approximate computation of several matrix products:

$$A^{\otimes N} \stackrel{\lambda}{\rightarrow} \bigoplus_h < m_h, n_h, p_h > .$$

Then we say

$$V_\tau(A) \geq \left(\sum_h (m_h n_h p_h)^\tau \right)^{1/N}.$$

We also symmetrize: if π is the cyclic permutation of the variables x,y,z in a trilinear form A, then we define

$$V_\tau(A) = (V_\tau(A \otimes \pi A \otimes \pi^2 A))^{1/3}.$$

It is immediate that "value" is *super-multiplicative:*

$$V_\tau(A \otimes B) \geq V_\tau(A) \times V_\tau(B),$$

where "$A \otimes B$" indicates the tensor product. "Value" is also *super-additive:*

$$V_\tau(A \oplus B) \geq V_\tau(A) + V_\tau(B),$$

where "$A \oplus B$" indicates that A and B share no variables. If A reduces to B then

$$V_\tau(A) \geq V_\tau(B).$$

For those familiar with Strassen's paper (1986), this notion of value V_τ is intermediate between Strassen's $\underset{\sim}{Q}$ and his $\underset{\sim}{R}$.

We will use this notion to analyze a more complicated version of our present construction, which will yield the exponent of 2.376.

We start with the tensor square of Construction (10), but we relabel the superscripts (regroup variables):

$$
\begin{aligned}
(q+2)^2 \to \sum_{i,k=1}^{q} \Big(& x^{[0]}_{0,0} y^{[2]}_{i,k} z^{[2]}_{i,k} + x^{[1]}_{0,k} y^{[1]}_{i,0} z^{[2]}_{i,k} + x^{[1]}_{0,k} y^{[2]}_{i,k} z^{[1]}_{i,0} + x^{[1]}_{i,0} y^{[1]}_{0,k} z^{[2]}_{i,k} + x^{[2]}_{i,k} y^{[0]}_{0,0} z^{[2]}_{i,k} + \\
& + x^{[2]}_{i,k} y^{[1]}_{0,k} z^{[1]}_{i,0} + x^{[1]}_{i,0} y^{[2]}_{i,k} z^{[1]}_{0,k} + x^{[2]}_{i,k} y^{[1]}_{i,0} z^{[1]}_{0,k} + x^{[2]}_{i,k} y^{[2]}_{i,k} z^{[0]}_{0,0} \Big) \\
+ \sum_{i=1}^{q} \Big(& x^{[2]}_{0,q+1} y^{[1]}_{i,0} z^{[1]}_{i,0} + x^{[0]}_{0,0} y^{[3]}_{i,q+1} z^{[1]}_{i,0} + x^{[0]}_{0,0} y^{[1]}_{i,0} z^{[3]}_{i,q+1} + \\
& + x^{[3]}_{i,q+1} y^{[0]}_{0,0} z^{[1]}_{i,0} + x^{[1]}_{i,0} y^{[2]}_{0,q+1} z^{[1]}_{i,0} + x^{[1]}_{i,0} y^{[0]}_{0,0} z^{[3]}_{i,q+1} + \\
& + x^{[3]}_{i,q+1} y^{[1]}_{i,0} z^{[0]}_{0,0} + x^{[1]}_{i,0} y^{[3]}_{i,q+1} z^{[0]}_{0,0} + x^{[1]}_{i,0} y^{[1]}_{i,0} z^{[2]}_{0,q+1} \Big) \\
+ \sum_{k=1}^{q} \Big(& x^{[2]}_{q+1,0} y^{[1]}_{0,k} z^{[1]}_{0,k} + x^{[3]}_{q+1,k} y^{[0]}_{0,0} z^{[1]}_{0,k} + x^{[3]}_{q+1,k} y^{[1]}_{0,k} z^{[0]}_{0,0} + \\
& + x^{[0]}_{0,0} y^{[3]}_{q+1,k} z^{[1]}_{0,k} + x^{[1]}_{0,k} y^{[2]}_{q+1,0} z^{[1]}_{0,k} + x^{[1]}_{0,k} y^{[3]}_{q+1,k} z^{[0]}_{0,0} + \\
& + x^{[0]}_{0,0} y^{[1]}_{0,k} z^{[3]}_{q+1,k} + x^{[1]}_{0,k} y^{[0]}_{0,0} z^{[3]}_{q+1,k} + x^{[1]}_{0,k} y^{[1]}_{0,k} z^{[2]}_{q+1,0} \Big) \\
+ \Big(& x^{[4]}_{q+1,q+1} y^{[0]}_{0,0} z^{[0]}_{0,0} + x^{[2]}_{q+1,0} y^{[2]}_{0,q+1} z^{[0]}_{0,0} + x^{[2]}_{q+1,0} y^{[0]}_{0,0} z^{[2]}_{0,q+1} + \\
& + x^{[2]}_{0,q+1} y^{[2]}_{q+1,0} z^{[0]}_{0,0} + x^{[0]}_{0,0} y^{[4]}_{q+1,q+1} z^{[0]}_{0,0} + x^{[0]}_{0,0} y^{[2]}_{q+1,0} z^{[2]}_{0,q+1} + \\
& + x^{[2]}_{0,q+1} y^{[0]}_{0,0} z^{[2]}_{q+1,0} + x^{[0]}_{0,0} y^{[2]}_{0,q+1} z^{[2]}_{q+1,0} + x^{[0]}_{0,0} y^{[0]}_{0,0} z^{[4]}_{q+1,q+1} \Big)
\end{aligned}
\tag{11}
$$

We have divided the $(q+2)^2$ x-variables into five blocks:

$$
\begin{aligned}
X^{[0]} &= \{x^{[0]}_{0,0}\} \\
X^{[1]} &= \{x^{[1]}_{i,0}, x^{[1]}_{0,k}\} \\
X^{[2]} &= \{x^{[2]}_{q+1,0}, x^{[2]}_{i,k}, x^{[2]}_{0,q+1}\} \\
X^{[3]} &= \{x^{[3]}_{q+1,k}, x^{[3]}_{i,q+1}\} \\
X^{[4]} &= \{x^{[4]}_{q+1,q+1}\}.
\end{aligned}
$$

Here i,k denote indices that range from 1 to q. In the original tensor square, the superscripts would have been 2-vectors: $x^{[1,0]}_{i,0}$. We have added the two elements to form a single superscript: $x^{[1]}_{i,0}$.

Notice that if $X^{[I]} Y^{[J]} Z^{[K]}$ appears in the trilinear form, then

$$I + J + K = 4.$$

The trilinear form can be written in block form as:

$$\sum_{I+J+K=4} X^{[I]}Y^{[J]}Z^{[K]}.$$

There are four types of terms in this trilinear form.

$$X^{[0]}Y^{[0]}Z^{[4]} = x_{0,0}^{[0]}y_{0,0}^{[0]}z_{q+1,q+1}^{[4]}. \tag{a}$$

This is a matrix product of size $<1,1,1>$, whose "value" is 1. There are three such terms: $X^{[0]}Y^{[0]}Z^{[4]}$, $X^{[0]}Y^{[4]}Z^{[0]}$, $X^{[4]}Y^{[0]}Z^{[0]}$.

$$X^{[0]}Y^{[1]}Z^{[3]} = \sum_{i=1}^{q} x_{0,0}^{[0]}y_{i,0}^{[1]}z_{i,q+1}^{[3]} + \sum_{k=1}^{q} x_{0,0}^{[0]}y_{0,k}^{[1]}z_{q+1,k}^{[3]}. \tag{b}$$

This is a matrix product of size $<1,1,2q>$ (i.e. a scalar $x_{0,0}^{[0]}$ times the vector $<y_{i,0}^{[1]}, y_{0,k}^{[1]}>$). Its "value" is $(2q)^{\tau}$. There are six such terms: $X^{[0]}Y^{[1]}Z^{[3]}$, $X^{[0]}Y^{[3]}Z^{[1]}$, $X^{[1]}Y^{[0]}Z^{[3]}$, $X^{[1]}Y^{[3]}Z^{[0]}$, $X^{[3]}Y^{[0]}Z^{[1]}$, $X^{[3]}Y^{[1]}Z^{[0]}$.

$$X^{[0]}Y^{[2]}Z^{[2]} = x_{0,0}^{[0]}y_{q+1,0}^{[2]}z_{0,q+1}^{[2]} + x_{0,0}^{[0]}y_{0,q+1}^{[2]}z_{q+1,0}^{[2]} + \sum_{i,k=1}^{q} x_{0,0}^{[0]}y_{i,k}^{[2]}z_{i,k}^{[2]}. \tag{c}$$

This is another matrix product, of size $<1,1,q^2+2>$, with "value" $(q^2+2)^{\tau}$. There are three such terms: $X^{[0]}Y^{[2]}Z^{[2]}$, $X^{[2]}Y^{[0]}Z^{[2]}$, $X^{[2]}Y^{[2]}Z^{[0]}$.

$$X^{[1]}Y^{[1]}Z^{[2]} = \sum_{i=1}^{q} x_{i,0}^{[1]}y_{i,0}^{[1]}z_{0,q+1}^{[2]} + \sum_{k=1}^{q} x_{0,k}^{[1]}y_{0,k}^{[1]}z_{q+1,0}^{[2]} + \\ + \sum_{i,k=1}^{q} x_{i,0}^{[1]}y_{0,k}^{[1]}z_{i,k}^{[2]} + \sum_{i,k=1}^{q} x_{0,k}^{[1]}y_{i,0}^{[1]}z_{i,k}^{[2]}. \tag{d}$$

There are three such terms: $X^{[1]}Y^{[1]}Z^{[2]}$, $X^{[1]}Y^{[2]}Z^{[1]}$, $X^{[2]}Y^{[1]}Z^{[1]}$. In the lemma at the end of this section, we will find that, if $q \geq 3$, the "value" of this term is at least $2^{2/3}q^{\tau}(q^{3\tau}+2)^{1/3}$.

Take the Nth tensor power of Construction (11), where N is divisible by 3. Let α_ℓ, $0 \le \ell \le 4$, be positive real numbers (determined below) such that

$$\sum_{\ell=0}^{4} \alpha_\ell = 1, \qquad \sum_{\ell=0}^{4} \ell\alpha_\ell = 4/3.$$

Let A_ℓ be integer approximations to $\alpha_\ell N$ such that

$$\sum_{\ell=0}^{4} A_\ell = N, \qquad \sum_{\ell=0}^{4} \ell A_\ell = 4N/3.$$

Retain only those blocks of variables $(X^{[I]}, Y^{[J]}, Z^{[K]})$ such that

$$\#\{j \mid 1 \le j \le N,\ \ I_j = \ell\} = A_\ell,$$

setting the others to zero, where as before I_j picks out the jth index position.

Let M'' be the number of nonzero triples $(X^{[I]}, Y^{[J]}, Z^{[K]})$ containing a given block $X^{[I]}$. We have

$$M'' = \sum_{\{\eta_{\ell,m,n}\}} \frac{\prod\limits_{0\le\ell\le4} A_\ell!}{\prod\limits_{\ell+m+n=4} \eta_{\ell,m,n}!}, \tag{12}$$

where $\{\eta_{\ell,m,n}\}$ range over partitions of N such that

$$\sum_{m,n} \eta_{\ell,m,n} = A_\ell$$
$$\sum_{\ell,n} \eta_{\ell,m,n} = A_m$$
$$\sum_{\ell,m} \eta_{\ell,m,n} = A_n$$

and the only nonzero values of $\eta_{\ell,m,n}$ occur with $\ell + m + n = 4$, $0 \le \ell,m,n \le 4$. The summand in (12) is maximized at

$$
\begin{aligned}
&\eta_{\ell,m,n} = \gamma_{\ell,m,n} \\
&\gamma_{0,0,4} = \gamma_{0,4,0} = \gamma_{4,0,0} = \hat{a} \\
&\gamma_{0,1,3} = \gamma_{0,3,1} = \gamma_{1,0,3} = \gamma_{1,3,0} = \gamma_{3,0,1} = \gamma_{3,1,0} = \hat{b} \\
&\gamma_{0,2,2} = \gamma_{2,0,2} = \gamma_{2,2,0} = \hat{c} \\
&\gamma_{1,1,2} = \gamma_{1,2,1} = \gamma_{2,1,1} = \hat{d} \\
&A_0 = 2\hat{a} + 2\hat{b} + \hat{c} \\
&A_1 = 2\hat{b} + 2\hat{d} \\
&A_2 = 2\hat{c} + \hat{d} \\
&A_3 = 2\hat{b} \\
&A_4 = \hat{a}.
\end{aligned} \tag{13}
$$

In later calculations we will approximate M'' by its largest term, times a polynomial N^p in N.

Set $M = 2M'' + 1$. Construct the Salem-Spencer set. Choose random weights w_j, $0 \le j \le N$. Compute the hash as before, with a minor change (4 replacing 2) in the definition on the z-indices:

$$
\begin{aligned}
b_X(I) &\equiv \sum_{j=1}^{N} I_j w_j \bmod M \\
b_Y(J) &\equiv w_0 + \sum_{j=1}^{N} J_j w_j \bmod M \\
b_Z(K) &\equiv \left[w_0 + \sum_{j=1}^{N} (4 - K_j) w_j \right] /2 \bmod M.
\end{aligned}
$$

Retain only those variables mapping into B, setting others to zero. As before, a nonzero triple $X^{[I]}Y^{[J]}Z^{[K]}$ remaining in the trilinear form will have $b_X(I) = b_Y(J) = b_Z(K) \in B$.

After the usual pruning, we have approximately

$$
\binom{N}{A_0, A_1, A_2, A_3, A_4}
$$

triples of blocks $(X^{[I]}, Y^{[J]}, Z^{[K]})$ remaining. For a good portion of these triples (at least a fraction N^{-p} of the total) the N indices $j = 1,2, \ldots ,N$ will contain about $\gamma_{\ell,m,n}$ instances of $X^{[\ell]}Y^{[m]}Z^{[n]}$. So the value of each triple of blocks is about

$$[1]^{3\hat{a}}[(2q^{\tau})]^{6\hat{b}}[(q^2+2)^{\tau}]^{3\hat{c}}[2^{2/3}q^{\tau}(q^{3\tau}+2)^{1/3}]^{3\hat{d}}$$
$$=(2q)^{6\tau\hat{b}}(q^2+2)^{3\tau\hat{c}}[4q^{3\tau}(q^{3\tau}+2)]^{\hat{d}}.$$

Thus our auxiliary equation is

$$(q+2)^{2N}\geq N^{-p}\binom{N}{A_0, A_1, A_2, A_3, A_4}(2q)^{6\tau\hat{b}}(q^2+2)^{3\tau\hat{c}}[4q^{3\tau}(q^{3\tau}+2)]^{\hat{d}}.$$

We want to choose $\hat{a}$, $\hat{b}$, $\hat{c}$, $\hat{d}$ to maximize the right-hand side, subject to

$$3\hat{a}+6\hat{b}+3\hat{c}+3\hat{d}=N,$$

and with A_ℓ defined by (13).

In fact, letting $\hat{a}=\bar{a}N$, $\hat{b}=\bar{b}N$, $\hat{c}=\bar{c}N$, $\hat{d}=\bar{d}N$, letting N grow and taking N^{th} roots, we get

$$(q+2)^2=\frac{(2q)^{6\tau\bar{b}}(q^2+2)^{3\tau\bar{c}}[4q^{3\tau}(q^{3\tau}+2)]^{\bar{d}}}{(2\bar{a}+2\bar{b}+\bar{c})^{2\bar{a}+2\bar{b}+\bar{c}}(2\bar{b}+2\bar{d})^{2\bar{b}+2\bar{d}}(2\bar{c}+\bar{d})^{2\bar{c}+\bar{d}}(2\bar{b})^{2\bar{b}}(\bar{a})^{\bar{a}}}.$$

We wish to minimize τ with respect to $\bar{a}$, $\bar{b}$, $\bar{c}$, $\bar{d}, q$, subject to

$$3\bar{a}+6\bar{b}+3\bar{c}+3\bar{d}=1$$
$$\bar{a},\bar{b},\bar{c},\bar{d}>0$$
$$q\geq 3,\quad q\in\mathbb{Z}.$$

We find that

$$\bar{a}=0.000233$$
$$\bar{b}=0.012506$$
$$\bar{c}=0.102546$$
$$\bar{d}=0.205542$$
$$q=6$$

gives an exponent of

$$\omega\leq 3\tau<2.375477.$$

DISCUSSION

Here we essentially set $w_{2j-1} = 2_{2j}$, when we grouped together into one block the variables $x_{i,0}^{[1]}, x_{0,k}^{[1]}$, and similar groupings for the other blocks. Other linear dependences among the w_j remain to be investigated, including:

$$w_{3j-2} = w_{3j-1} = w_{3j}$$

or

$$w_{2j-1} = -2w_{2j}.$$

PROOF OF THE LEMMA

Lemma. *Assume $q \geq 3$. The "value" V_τ of the trilinear form*

$$\sum_{i=1}^{q} x_{i,0}^{[1,0]} y_{i,0}^{[1,0]} z_{0,q+1}^{[0,2]} + \sum_{k=1}^{q} x_{0,k}^{[0,1]} y_{0,k}^{[0,1]} z_{q+1,0}^{[2,0]} +$$
$$+ \sum_{i,k=1}^{q} x_{i,0}^{[1,0]} y_{0,k}^{[0,1]} z_{i,k}^{[1,1]} + \sum_{i,k=1}^{q} x_{0,k}^{[0,1]} y_{i,0}^{[1,0]} z_{i,k}^{[1,1]}$$

is at least $2^{2/3} q^{\tau} (q^{3\tau} + 2)^{1/3}$.

Note: The superscripts are different from those in the application of this lemma, but they are still uniquely determined by the subscripts.

Proof. Take the $2N^{th}$ tensor power. Retain only those X-blocks with exactly N indices of [1,0] and N of [0,1]. Retain only those Y-blocks with exactly N indices of [1,0] and N of [0,1]. Retain those Z-blocks with exactly L indices of [2,0], exactly L of [0,2], and $2G = 2N - 2L$ of [1,1], where

$$L = \left[\frac{2}{q^{3\tau} + 2} N \right] \text{ and } G = N - L.$$

The number of X-blocks is $\binom{2N}{N}$. The number of nonzero triples $(X^{[I]} Y^{[J]} Z^{[K]})$ containing a given X-block is

$$\binom{L+G}{G}^2.$$

The same numbers hold for Y-blocks, but not Z-blocks; see below.

Set

$$M = 4\binom{L+G}{G}^2 + 1.$$

Construct a Salem-Spencer set B, define a hash as before, and set to zero blocks not hashing into B. Eliminate instances where an X-block or a Y-block is shared among two or more nonzero triples.

There are $\binom{2N}{L,L,2G}$ Z-blocks, and to each there are initially $\binom{2G}{G}$ triples $(X^{[I]}Y^{[J]}Z^{[K]})$. Counting only those with $b_X(I) = b_Y(J) = b_Z(K)$ we have about

$$\binom{2G}{G}M^{-1} \simeq \frac{1}{4}\binom{2G}{G}\binom{L+G}{G}^{-2} \equiv H$$

A calculation shows that $H >> 1$ if $G/L > 3.41$.

This is, in Strassen's terminology, a C-tensor over $< 1,H,1 >$, where C is the class of tensors consisting of matrix products $< m,n,p >$ with $mnp = (q^2)^{2G}(q)^{2L} = q^{4G+2L}$.

In short, we have $\binom{2L+2G}{L,L,2G}$ disjoint objects, each of which is a C-tensor over $< 1,H,1 >$. The auxiliary equation is:

$$V_\tau^{2N} \simeq \binom{2N}{L,L,2G}\left[\frac{\binom{2G}{G}}{4\binom{N}{G}^2}\right]^{2/3}(q^{4G+2L})^\tau$$

$$\simeq \frac{(2N)^{2N}}{L^L L^L (2G)^{2G}}\left[\frac{(2G)^{2G}}{G^G G^G}\;\frac{G^G L^L G^G L^L}{4N^N N^N}\right]^{2/3} q^{(4G+2L)\tau}$$

$$V_\tau^{3N} \simeq \frac{(2N)^{3N}}{L^L (2G)^G N^{2N}}\, q^{(6G+3L)\tau} \simeq \frac{2^{2N+L}N^N}{L^L G^G}\, q^{3N\tau} q^{3G\tau}$$

$$\simeq 2^{2N} q^{3N\tau}\binom{N}{L} 2^L (q^{3\tau})^G.$$

Selecting

$$\frac{L}{L+G} = \frac{2}{q^{3\tau}+2}, \quad \frac{G}{L+G} = \frac{q^{3\tau}}{q^{3\tau}+2}$$

maximizes this estimate at

$$2^{2N}q^{3N\tau}(q^{3\tau}+2)^{N}$$

and, taking $3N^{th}$ roots with N large,

$$V_\tau \geq 2^{2/3}q^{\tau}(q^{3\tau}+2)^{1/3}.$$

Finally, the estimate

$$\frac{G}{L}=\frac{q^{3\tau}}{2}>3.41$$

is ensured if $q \geq 3$, since $3\tau \geq 2$.

9. Related schemes

We developed a few other techniques on the way to the present paper. While these did not lead to a lower matrix exponent, they may be of interest when applied some other starting algorithm.

ASYMMETRIC VERSION

The first technique we present is an asymmetric version of the technique presented in the previous two sections. We start with an asymmetric basic algorithm, with a block structure and a fine structure. After a suitable tensor power, instead of pulling out several block scalar products (whose fine structure is that of matrix products), we will instead pull out several block vector products. In Strassen's terminology, we will be developing several C-tensors over $<1,H,1>$ for some large value of H. This is made possible by the Salem-Spencer Theorem, as before. We then use Strassen's machinery to estimate the value of these C-tensors over $<1,H,1>$.

We illustrate with a basic algorithm closely related to Strassen's. It uses $q+1$ multiplications.

$$\begin{aligned}&\sum_{i=1}^{q}(x_0^{[0]}+\lambda x_i^{[1]})(y_0^{[0]}+\lambda y_i^{[1]})(z_i^{[1]}\lambda^{-1})+(x_0^{[0]})(y_0^{[0]})(z_{q+1}^{[2]}-\sum z_i^{[1]}\lambda^{-1})\\&=\sum_{i=1}^{q}(x_i^{[1]}y_0^{[0]}z_i^{[1]}+x_0^{[0]}y_i^{[1]}z_i^{[1]})+x_0^{[0]}y_0^{[0]}z_{q+1}^{[2]}+O(\lambda).\end{aligned} \tag{14}$$

Later we will determine real numbers $\alpha,\beta>0$ with $2\alpha+\beta=1$. Choose ε small and N "large enough". Set

$$L = [\alpha N],\ G = N - 2L,\ M = 2\binom{L+G}{L} + 1.$$

Construct a Salem-Spencer set B using this value of M.

Take the Nth tensor power of Construction (10). Set to zero all blocks except: X - or Y-blocks having exactly $G + L$ indices of 0 and L indices of 1, and Z-blocks having exactly G indices of 2 and $2L$ indices of 1. Select random integers $w_0, w_1, \ldots, w_N$ mod M, compute the hash as before, and set to zero any variable not hashing to an element of B. Then any nonzero triple of blocks $(X^{[I]}, Y^{[J]}, Z^{[K]})$ whose product appears in the construction has, as before,

$$b_X(I) = b_Y(J) = b_Z(K) \in B.$$

Any X-block X_I is compatible with exactly $\binom{L+G}{L}$ Z-blocks Z_K: the L indices of 1 in I all correspond to indices of 1 in K, while of the $G + L$ indices of 0 in I, exactly G correspond to indices of 2 in K, and L to indices of 1. Thus the expected number of Z-blocks such that $b_Z(K) = b_X(I)$ is about 1/2. If there are more than two such Z-blocks, zero a corresponding Y-block. Similarly, if a Y-block is involved in at least two triples $(X^{[I]}, Y^{[J]}, Z^{[K]})$,$(X^{[I']}, Y^{[J]}, Z^{[K']})$, we zero one of the X-blocks. Thus each X- or Y-block is involved in at most one triple $(X^{[I]}, Y^{[J]}, Z^{[K]})$ after this pruning.

We do not make such a requirement on the Z-blocks; we allow each Z-block to be involved in a large number of triples:

$$(X^{[I_h]}, Y^{[J_h]}, Z^{[K]}),\quad 1 \le h \le H.$$

For a given Z-block, the expected number of such triples (before pruning) is

$$\mu = \binom{2L}{L} M^{-1}.$$

As before, given two blocks $X^{[I]}, X^{[I']}$ compatible with a given block Z_K, the events that $b_X(I) = b_Z(K)$ and $b_X(I') = b_Z(K)$ are (pairwise) independent, so that we can estimate the variance in the number of triples (before pruning) as

$$\sigma^2 = \binom{2L}{L}(M^{-1} - M^{-2}).$$

Then Chebyshev's inequality can be applied to say that

$$\text{Prob}(\#blocks < \mu - 3\sigma) \le \frac{1}{10}.$$

For a given block $Z^{[K]}$ with $b_Z(K) \in B$, the probability of at least $\mu - 3\sigma$ triples $(X^{[I_h]}, Y^{[J_h]}, Z^{[K]})$ before pruning (i.e. $b_X(I_h) = b_Y(J_h) = b_Z(K)$) is at least 9/10. Pruning eliminates only a constant fraction of the triples. Together with the observation that $\sigma << \mu$, we obtain:

Lemma: *There are constants* c_1, c_2 *such that, averaging over the choice of* w_j, *the expected number of blocks* $Z^{[K]}$ *with at least* $c_1 \binom{2L}{L} M^{-1}$ *associated triples* $(X^{[I_h]}, Y^{[J_h]}, Z^{[K]})$ *remaining after pruning, is at least* $c_2 M' M^{-1} \binom{2L+G}{G}$.

That is, we will have at least $c_2 M' M^{-1} \binom{2L+G}{G}$ independent objects, each of which is a C-tensor over $< 1, c_1 \binom{2L}{L} M^{-1}, 1 >$, where C is the class of matrix multiplication problems $< m, n, p >$ with $mnp = q^{2L}$.

Applying Strassen's formula, we get

$$\omega \le 3\tau_N, \quad (q+1)^N \ge c_2 M' M^{-1} \binom{2L+G}{G} \left[c_1 \binom{2L}{L} M^{-1} \right]^{2/3} q^{(2L\tau_N)}.$$

Letting ε shrink and N grow, and taking N^{th} roots,

$$\omega \le 3\tau, \quad (q+1) \ge (2\alpha)^{-2\alpha} \beta^{-\beta} \left[2^{2\alpha} \frac{\alpha^\alpha \beta^\beta}{(\alpha+\beta)^{\alpha+\beta}} \right]^{2/3} q^{2\alpha\tau}.$$

(As a consistency check, note that the limiting case, $\alpha = 0.5$, $\beta = 0$, recovers Strassen's formula $q + 1 = 2^{2/3} q^\tau$, by disallowing terms containing z_{q+1}.) Selecting $q = 4$ and $\alpha = 0.485$, we optimize ω at 2.4602. This exponent is not as good as that obtained in the previous section, but the techniques exposed here may be more widely applicable.

"STRASSEN SQUARED"

Another possibility is to iterate Strassen's construction. Where Strassen develops a C-tensor over $< 1,k,1 >$, where C is a class of matrix products $< m,n,p >$ for a fixed value of mnp, we instead develop a D-tensor over $< 1,1,k >$, where D is a class of C-tensors over $< 1, k',1 >$, with C again a class of matrix products.

To illustrate, start with Construction (5) from Section 5, using $q + 2$ multiplications to obtain

$$\sum_{i=1}^{q} (x_0 y_i z_i + x_i y_0 z_i + x_i y_i z_0) + O(\lambda).$$

Replace λ by λ^2 throughout. Take the tensor product of this construction with itself; $(q+2)^2$ multiplications yield

$$\sum_{i=1}^{q}\sum_{j=1}^{q}(x_{0,0}y_{i,j}z_{i,j}+x_{0,j}y_{i,0}z_{i,j}+x_{0,j}y_{i,j}z_{i,0}+x_{i,0}y_{0,j}z_{i,j}+x_{i,j}y_{0,0}z_{i,j}$$
$$+x_{i,j}y_{0,j}z_{i,0}+x_{i,0}y_{i,j}z_{0,j}+x_{i,j}y_{i,0}z_{0,j}+x_{i,j}y_{i,j}z_{0,0})+O(\lambda^2).$$

Set $x_{0,0}$ to 0. Multiply $x_{i,j}$ by λ, and multiply $y_{0,0}$ and $z_{0,0}$ by λ^{-1}. This kills three terms:

$$x_{0,0}y_{i,j}z_{i,j},\quad x_{i,j}y_{0,j}z_{i,0},\quad x_{i,j}y_{i,0}z_{0,j}.$$

We are left with

$$\sum_{i=1}^{q}\sum_{j=1}^{q}[(x_{0,j}y_{i,0}z_{i,j}+x_{i,0}y_{0,j}z_{i,j}+x_{i,j}y_{0,0}z_{i,j})+(x_{0,j}y_{i,j}z_{i,0}+x_{i,0}y_{i,j}z_{0,j}+x_{i,j}y_{i,j}z_{0,0})]+O(\lambda).$$

The first parenthesized expression has the form of a C-tensor over $<1,3,1>$, where C is the class of matrix products $<m,n,p>$ with $mnp=q^2$; the second is a C-tensor over $<3,1,1>$. Let D denote the class of C-tensors over $<m',n',p'>$ with $m'n'p'=3$. The two parenthesized expressions share x-variables, and they fit together as a D-tensor over $<1,1,2>$. By Strassen's machinery, we can write:

$$\omega \le 3\tau,\quad (q+2)^2 = 2^{2/3}3^{2/3}q^{2\tau}$$
$$\omega \le \log_q\left(\frac{(q+2)^3}{6}\right).$$

Setting $q=9$, we get $\omega \le \log_9(11^3/6) < 2.459$. Again, this is not as profitable as the development in Section 6, where the corresponding equation was

$$\omega \le \log_q\left(\frac{(q+2)^3}{6.75}\right).$$

10. Remarks

We are applying the τ-theorem in a special case, namely when all the matrix products are of the same shape. Examining the proof of the τ-theorem, one finds that this eliminates some of the sources of inefficiency for matrices of moderate size. Also, by altering the construction slightly, we can eliminate the use of λ, another contributor to the inefficiency inherent in the τ-theorem.

Nonetheless, the large values of M dictated by the Salem-Spencer theorem still make the present algorithm wildly impractical for any conceivable applications.

A remarkable feature of the present approach is that we can use basic constructions in which the number of x-variables is equal to the number of multiplications, as is the number of y- or z-variables. (In Section 7, this number is $q+2$.) Before Strassen's 1986 paper, this was not possible. Coppersmith and Winograd (1982) had shown that for any basic algorithm (approximately) computing (several) matrix products, the number of multiplications had to strictly exceed the number of x -variables (or else y- or z-variables), except in trivial cases. This had led to the acceleration techniques of Coppersmith and Winograd (1982), but had also implied that one could never achieve $\omega = 2$ by starting with a *fixed, finite-size basic algorithm*. But now Strassen's techniques have removed the necessity of starting with a matrix product, and, coupled with the techniques of the present paper, allow the hope of someday achieving $\omega = 2$. (See the following section.)

An open question is a possible analogue of the Salem-Spencer theorem in characteristic two: For arbitrary $\varepsilon > 0$ and large enough N, do there exist subsets A, B, C of $(\mathbb{Z}_2)^N$ of size $|A| = |B| = |C| = (2^N)^{1-\varepsilon}$, such that each element $a \in A$, $b \in B$, or $c \in C$ is involved in exactly one relation of the form

$$a + b + c = 0?$$

That is,

$$\begin{aligned}
&\forall a \in A \ \exists! b \in B,\ c \in C: \quad a + b + c = 0 \\
&\forall b \in B \ \exists! a \in A,\ c \in C: \quad a + b + c = 0 \\
&\forall c \in C \ \exists! a \in A,\ b \in B: \quad a + b + c = 0
\end{aligned}$$

With such a construction, the present techniques would become applicable to a wider class of starting algorithms. For example, our first algorithm involved three block products, $X^{[0]}Y^{[1]}Z^{[1]}$, $X^{[1]}Y^{[0]}Z^{[1]}$, $X^{[1]}Y^{[1]}Z^{[0]}$, so that if $X^{[I]}Y^{[J]}Z^{[K]}$ appeared we knew $I + J + K = 2$. If in addition the block product $X^{[0]}Y^{[0]}Z^{[0]}$ appeared, our condition on the indices would become $I + J + K = 0 \in \mathbb{Z}_2$. A characteristic-two version of the Salem-Spencer theorem would become useful in that situation.

11. Can We Achieve $\omega = 2$?

In closing, we sketch the relation between a hypothetical combinatorial construction and the elusive

$$\omega \stackrel{?}{=} 2.$$

Definition. An Abelian group G (with at least two elements) and a subset S of G satisfy the *no three disjoint equivoluminous subsets* property if: whenever T_1, T_2, T_3 are three disjoint subsets of S, not all empty, they cannot all have the same sum in G:

$$\sum_{s \in T_1} s \neq \sum_{s \in T_2} s \quad \text{or} \quad \sum_{s \in T_2} s \neq \sum_{s \in T_3} s.$$

Assume for now that we can find a sequence of pairs G,S with the no three disjoint equivoluminous subsets property, such that $(\log_2|G|)/|S|$ approaches 0.

If the trilinear form

$$x_0y_1z_2 + x_0y_2z_1 + x_1y_0z_2 + x_2y_0z_1 + x_1y_2z_0 + x_2y_1z_0$$

had border rank 3, then by the techniques of Chapter 6 (easy case) with $q = 2$, we could prove $\omega = 2$. Unfortunately, this form has border rank 4. But if the nth tensor power of this form had border rank $3^{n+o(n)}$, the same techniques would still yield $\omega = 2$. This is our goal.

We will use a Fourier transform to get a related form to have rank 3. This related form has extra terms, which we must cancel somehow. We will multiply variables by roots of unity, take a large tensor power, and take a Fourier transform again, and use the no three disjoint equivoluminous subsets property to arrange the cancellation of the unwanted terms. This will show that the nth tensor power has rank $3^{n+o(n)}$. Then we will apply the techniques from this paper to derive $\omega = 2$, given the existence of (G,S) with $(\log_2|G|)/|S|$ approaching 0.

Let $S = \{s_1, s_2, ..., s_n\}$ and G satisfy the no three disjoint equivoluminous subsets property, with $\varepsilon = (\log_2|G|)/|S|$.

Let χ and ψ be two independent characters on G. Select i from $\{1, 2, ..., n\}$. Evaluate the characters $\chi(s_i)$, $\psi(s_i)$, which are complex numbers of magnitude 1. Select $r = r(\chi, \psi, s_i)$ to satisfy $r^3 = \chi(s_i)/\psi(s_i)$.

Evaluate the following sum, where ω_i ranges over the three complex cube roots of 1:

$$U_i \equiv \sum_{\omega_i} \frac{1}{3} \left(x_0 + \chi(s_i)^{-1} r\omega_i x_1 + r^{-1}\omega_i^2 x_2\right) \times$$
$$\times \left(\chi(s_i) r^{-1} y_0 + \omega_i y_1 + \psi(s_i) r\omega_i^2 y_2\right) \times$$
$$\times \left(r z_0 + \psi(s_i)^{-1} r^{-1} \omega_i z_1 + \omega_i^2 z_2\right)$$

We see that this has trilinear rank at most 3. By multiplying by powers of ω_i and summing over ω_i, we are taking a three-point Fourier transform, and the terms which survive in the product are those for which the exponent of ω_i is divisible by 3:

$$U_i = (x_0 y_1 z_2 + x_0 y_2 z_1 + x_1 y_0 z_2 + x_2 y_0 z_1 + x_1 y_2 z_0 + x_2 y_1 z_0) +$$
$$+ (\chi(s_i) x_0 y_0 z_0) + (\chi(s_i)^{-1} \psi(s_i)^{-1} x_1 y_1 z_1) + (\psi(s_i) x_2 y_2 z_2).$$

We may think of this as

$$U_i = (\mathrm{Good}_i) + \chi(s_i)\mathrm{Bad1}_i + \chi(s_i)^{-1}\psi(s_i)^{-1}\mathrm{Bad2}_i + \psi(s_i)\mathrm{Bad3}_i,$$

where Good_i is the sum ot the first six terms (which we want), and $\mathrm{Bad}\ell_i$ are the unwanted terms.

Now take the *tensor product* of the various U_i:

$$\bigotimes_{1 \le i \le n} U_i = \bigotimes_i \left((\mathrm{Good}_i) + \chi(s_i)\mathrm{Bad1}_i + \chi(s_i)^{-1}\psi(s_i)^{-1}\mathrm{Bad2}_i + \psi(s_i)\mathrm{Bad3}_i \right).$$

This is the sum of 4^n terms, each a tensor product of trilinear forms of the sort Good_i or $\mathrm{Bad}\ell_i$. The first term,

$$\mathrm{GOOD} = \bigotimes_i \mathrm{Good}_i$$

is the one we want. For each of the other $4^n - 1$ terms, for $\ell = 1, 2, 3$, let T_ℓ be the set of s_i corresponding to those indices i for which $\mathrm{Bad}\ell_i$ is included in the product. The coefficient of this term is

$$\left(\prod_{i \in T_1} \chi(s_i)\right) \left(\prod_{i \in T_2} \chi(s_i)^{-1}\psi(s_i)^{-1}\right) \left(\prod_{i \in T_3} \psi(s_i)\right)$$
$$= \chi\left(\sum_{i \in T_1} s_i - \sum_{i \in T_2} s_i\right) \times \psi\left(-\sum_{i \in T_2} s_i + \sum_{i \in T_3} s_i\right)$$

Since the T_ℓ are disjoint and not all empty, the no three disjoint equivoluminous subsets property implies that either the argument of χ or the argument of ψ is not the identity element of G.

Now sum over the various choices of χ and ψ. The "GOOD" term will accumulate (and get multiplied by $|G|^2$, which we can deal with), while all the other terms will vanish, because we are summing over all characters χ (or ψ) at an element of G other than the identity.

This shows that the trilinear rank of

$$\bigotimes_{1\le i\le n} \mathrm{Good}_i = \bigotimes_{1\le i\le n} (x_0y_1z_2 + x_0y_2z_1 + x_1y_0z_2 + x_2y_0z_1 + x_1y_2z_0 + x_2y_1z_0)$$

is no more than $|G|^2 3^n$.

Note here that, if necessary, we can replace G by a Cartesian product of G with itself N times, and replace S by N copies of S, one in each copy of G. A typical element of the large S will be $(0,\ldots,0,\, s_i,\, 0,\ldots,0)$. Then $\varepsilon = (\log_2|G|)/|S|$ remains unchanged, and the no three disjoint equivoluminous subsets property is inherited.

Following the arguments from Chapter 6, we would then obtain, by analogy to Equation (9),

$$\begin{gathered} 2^{6\varepsilon}3^3 \ge (27/4)2^{3\tau} \\ \omega = 3\tau \le \log_2(2^{6\varepsilon} \times 4) = 2 + 6\varepsilon. \end{gathered} \tag{15}$$

With the assumption that $\varepsilon = (\log_2|G|)/|S|$ approaches 0, this would yield $\omega = 2$.

We have not been able to determine whether there exist such pairs (G,S) with $(\log_2|G|)/|S|$ approaching 0.

References.

Behrend, F.A. (1946). On sets of integers which contain no three terms in arithmetical progression. *Proc. Nat. Acad. Sci. USA* **32**, 331-332.

Coppersmith, D., Winograd, S. (1982). On the Asymptotic Complexity of Matrix Multiplication. *SIAM Journal on Computing,* Vol. 11, No. 3, 472-492.

Coppersmith, D., Winograd, S. (1986). Matrix Multiplication via Behrend's Theorem. *Research Report RC 12104,* IBM T.J. Watson Research Center, Yorktown Heights, N.Y., 10598, August 29, 1986.

Coppersmith, D., Winograd, S. (1987). Matrix Multiplication via Arithmetic Progressions. *Proc. 19th Ann. ACM Symp. on Theory of Computing,* 1-6.

Pan, V.Ya. (1978). Strassen Algorithm Is Not Optimal. Trilinear Technique of Aggregating, Uniting and Canceling for Constructing Fast Algorithms for Matrix Multiplication. *Proc. 19th Ann. IEEE Symp. on Foundations of Computer Science,* 166-176.

Pan, V.Ya. (1984). How to Multiply Matrices Faster. *Springer Lecture Notes in Computer Science,* vol. 179.

Schönhage, A. (1981). Partial and Total Matrix Multiplication. *SIAM J. on Computing,* 10, 3, 434-456.

Salem, R., Spencer, D.C. (1942). On sets of integers which contain no three terms in arithmetical progression. *Proc. Nat. Acad. Sci. USA* **28**, 561-563.

Strassen, V. (1986). The Asymptotic Spectrum of Tensors and the Exponent of Matrix Multiplication. *Proc. 27th Ann. IEEE Symp. on Foundations of Computer Science,* 49-54.

Functional Decomposition of Polynomials: the Tame Case

JOACHIM VON ZUR GATHEN

Department of Computer Science, University of Toronto
Toronto, Ontario M5S 1A4, Canada

(*Received* 22 *June* 1988)

If g and h are polynomials of degrees r and s over a field, their functional composition $f = g(h)$ has degree $n = rs$. The functional decomposition problem is: given f of degree $n = rs$, determine whether such g and h exist, and, in the affirmative case, compute them. We first deal with univariate polynomials, and present sequential algorithms that use $O(n \log^2 n \log\log n)$ arithmetic operations, and a parallel algorithm with optimal depth $O(\log n)$. Then we consider the case where f and h are multivariate, and g is univariate. All algorithms work only in the "tame" case, where the characteristic of the field does not divide r.

1. Introduction

If F is a field and $g, h \in F[x]$, then $f = g \circ h = g(h) \in F[x]$ is their (functional) *composition*, and (g, h) is a (functional) *decomposition* of f. Given $f \in F[x]$, there exists an essentially unique *complete* decomposition $f = f_1 \circ f_2 \circ \cdots \circ f_k$, where $f_1, \cdots, f_k \in F[x]$ are indecomposable polynomials (we give references in Section 2). This result is valid if the characteristic p of F does not divide the degree of f.

We start with the following decomposition problem: given $f \in F[x]$ of degree n, and $r, s \in \mathbb{N}$ with $n = rs$, decide whether there exist $g, h \in F[x]$ of degrees r, s, respectively, such that $f = g \circ h$. For some time, this problem was considered to be computationally hard: the security of a cryptographic protocol was based on its hardness (Cade 1985, broken by Berkovits and Lidl & Niederreiter), and exponential-time algorithms (in characteristic zero) were given by Alagar & Thanh (1985) and Barton & Zippel (1985) (a first version of which appeared in 1976). Major progress was made by Kozen & Landau (1989), who presented the first polynomial-time algorithm, in 1986. Their algorithm runs in sequential time $O(n^2)$, and $O(\log^2 n)$ in parallel. Gutiérrez *et al.* (1989) present a similar algorithm.

The present paper continues the work of Kozen & Landau in several directions, using their basic method: faster sequential and parallel algorithms, Boolean computations,

This work was supported by National Science and Engineering Council of Canada, grant A2514, by DFG grant SFB 124, and by Fundación Andes, beca C-10246. Parts of this work were done during visits to Universität des Saarlandes, Saarbrücken, Germany, and to Pontificia Universidad Católica de Chile, Santiago, Chile, and as a Visiting Fellow at the Computer Sciences Laboratory, Australian National University, Canberra, Australia.

complete decompositions, and multivariate polynomials. A fast decomposition method in Section 2 uses $O(n \log^2 n \log\log n)$ arithmetic operations, and $O(n \log^2 n)$ if F supports a Fast Fourier Transform. We find a complete decomposition with $O(n^{1+\epsilon})$ operations, for any $\epsilon > 0$. Over $\mathbb{Q}$, our approach yields a polynomial bound on the binary length of intermediate results, and thus shows that the decision problem is in the Boolean complexity class P. A parallel algorithm of optimal (up to constant factors) depth $O(\log n)$ is in Section 3.

As an application, Section 4 gives a (random) polynomial-time algorithm to decide whether a *separated polynomial* $f_1(x) - f_2(y) \in F[x,y]$ has a nontrivial separated factor, assuming that F supports a (random) polynomial-time root-finding procedure.

In Section 5 we consider decompositions of the form $f = g \circ h$ with $f, h \in F[x_1, \ldots, x_m]$ and $g \in F[x]$. The first polynomial-time algorithm for this problem is in Dickerson (1987). We present a conceptually simple Newton approach that yields polynomial-time algorithms for densely presented inputs, and random polynomial time for inputs given by arithmetic circuits. For the important sparse representation, we have no polynomial-time results.

All results of this paper work only in the "tame" case where $p = \text{char}(F)$ does not divide r. A subsequent paper (von zur Gathen 1988) deals with the "wild" case, where p divides r; Kozen & Landau (1989) also have results for that case. (The terminology of "tame" and "wild" is borrowed from number theory, regarding r as some "ramification index"; see e.g., Hasse 1980.) Some of the present results were reported in von zur Gathen, Kozen & Landau (1987).

2. Fast univariate decomposition

The subject of this and the next section is the following decomposition problem $\text{DEC}^F_{n,r}$. We have a field F, integers $n, r \in \mathbb{N}$ with r dividing n, and $f \in F[x]$ of degree n. Let $s = n/r$. The problem is to decide whether there exist $g, h \in F[x]$ of degrees r, s, respectively, such that $f = g \circ h = g(h)$ is the composition of g with h, and, in the affirmative case, to compute g and h. f is *indecomposable* if no such g and h exist, with $2 \le r < n$. The "tame" case is when the characteristic p of F does not divide r. This paper deals only with the tame case.

For the question of uniqueness, note the following three types of ambiguous decompositions. For any $u \in F[x]$, $c, d \in F$, $c \neq 0$, and $r, m \geq 2$ we have $u \circ (cx+d) \circ ((x-d)/c) = u$, $(x^m \cdot u^r) \circ x^r = x^r \circ (x^m \cdot u(x^r))$, and $T_r \circ T_m = T_m \circ T_r (= T_{rm})$, where T_i is the ith Chebyshev polynomial. "Ritt's First Theorem" states that a *complete decomposition* $f = f_1 \circ \cdots \circ f_k$ with $f_1, \ldots, f_k$ indecomposable is unique up to these ambiguities, i.e., that any two complete decompositions can be obtained from each other using these equalities (Ritt (1922) for $F = \mathbb{C}$, Engstrom (1941) for $p = 0$, Fried & MacRae (1969a) for $p = 0$ or $p > n$).

Decompositions are intimately related to the intermediate fields between $F(f)$ and $F(x)$ (Ritt 1922, Dorey & Whaples 1974) and between F and a splitting field of f over F (Kozen & Landau 1989).

If $f = g \circ h$ and a and c are the leading coefficients of f and h, respectively, then an affine linear transformation yields

$$\frac{f}{a} = \left(\frac{1}{a} g(cx + h(0))\right) \circ \frac{h - h(0)}{c},$$

a *normal* decomposition of a monic polynomial into monic polynomials, where the second composition factor has constant term zero. So we can assume that f, g, and h are monic, and that $h(0) = 0$. Denoting by $\mathcal{M} \subseteq F[x]$ the set of monic polynomials, we consider the relation of *normal decompositions*

$$\begin{aligned} \mathrm{DEC}_{n,r}^F &= \{(f,(g,h)) \in \mathcal{M} \times \mathcal{M}^2 : \\ & \quad f = g \circ h,\ \deg f = n,\ \deg g = r, \text{ and } h(0) = 0\}. \end{aligned}$$

Formally, the computational problem has $f \in \mathcal{M}$ and $r \in \mathbb{N}$ as input, and as output the set of all $(g, h) \in \mathcal{M}^2$ with $(f,(g,h)) \in \mathrm{DEC}_{n,r}^F$.

In the tame case, the algorithm of Kozen & Landau (1989) (or Corollary 2.3 below) implies that for every f there exists at most one such (g, h), so that we can view $\mathrm{DEC}_{n,r}^F : \mathcal{M} \longrightarrow \mathcal{M}^2$ as a partial function. Furthermore, the problem is rational, i.e., if $f \in F[x]$ and there exists a field extension $K \supseteq F$ and $(f,(g,h)) \in \mathrm{DEC}_{n,r}^K$, then in fact $g, h \in F[x]$ (Levi 1942). Both facts and Ritt's First Theorem may fail in the "wild" case, where p divides r. Incidentally, a variant of the new algorithms is implicit in Levi (1942), Section 2.

Brent & Kung (1978) deal with a different problem: given $g, h \in F[x]$ of degree at most n, compute the lowest n coefficients of $g \circ h$, and, assuming $g(0) = 0$ and $g'(0) = 1$, of the functional inverse of g. They give algorithms using time $O((n \log n)^{3/2})$, if F supports a Fast Fourier Transform. Ritzmann (1984) has a better bound for computing such compositions, in a different "numerical" model.

Our model of computation is the "arithmetic Boolean circuit", which uses indeterminate inputs (in our case the coefficients of f), constants from F, the arithmetic operations $+, -, *, /$, tests $a \stackrel{?}{=} 0$, binary Boolean operations on the resulting Boolean values, and selection gates that select one of two arithmetic values according to the value of a Boolean third input (see von zur Gathen 1986). In our analyses, we usually neglect the Boolean cost, since it is always dominated by the arithmetic cost.

In Theorem 2.7, we consider the ground field $\mathbb{Q}$ and inputs presented in binary and Boolean computations, say on a Turing machine or on Boolean circuits, and derive a polynomial bound on the binary length of intermediate results.

We will use the following well-known facts about some computational problems. Let $M = M_F : \mathbb{N} \longrightarrow \mathbb{R}$ be such that the product of two polynomials in $F[x]$ of degree at most n can be computed with $O(M(n))$ arithmetic operations. We can choose $M(n) = n \log n \log\log n$ (Schönhage 1977, Cantor & Kaltofen 1987), and $M(n) = n \log n$ if F supports a Fast Fourier Transform.

FACT 2.1. *(i)* [Inversion] *Given* $f \in F[x]$ *with* $f(0) = 1$, *one can compute* $f^{-1} \bmod x^{n+1}$ *with* $O(M(n))$ *operations.*

(ii) [Division with remainder] *Given* $f, g \in F[x]$ *with degree at most* n *and* $g \neq 0$, *one can compute* $q, r \in F[x]$ *such that* $f = qg + r$ *and* $\deg r < \deg g$ *with* $O(M(n))$ *operations.*

(iii) [Roots] *Given* $f \in F[x]$ *of degree* n *with* $f(0) = 1$, *and* $r \in \mathbb{N}$ *not divisible by* $\mathrm{char}(F)$, *one can compute the unique* $h \in F[x]$ *of degree at most* n *such that* $h^r \equiv f \bmod x^{n+1}$ *and* $h(0) = 1$ *with* $O(M(n) \log r)$ *operations.*

(iv) [Taylor expansion] *Given* $f, h \in F[x]$ *of degrees* n, s *respectively, let* $r = \lceil n/s \rceil$. *One can compute with* $O(M(n) \log n)$ *operations the unique* $b_0, \ldots, b_r$ *in* $F[x]$ *such that*

$$f = \sum_{0 \le i \le r} b_i h^i \text{ and } \deg b_i < \deg h \text{ for all } i.$$

(v) [Composition] *Given $g, h \in F[x]$ of degrees r, s, repectively, one can compute $g \circ h$ in $O(M(n) \log n)$ operations, where $n = rs$.*

PROOF. (i) and (ii) are in Borodin & Munro (1975), Section 4.4. (iii) follows with Newton iteration, as in Brent & Kung (1978). The generalized "Taylor expansion" (iv) of f around h is most familiar when $h = x - h_0$ is linear, and we have the usual Taylor expansion of f around h_0. A divide-and-conquer approach gives the following algorithm. Let $t = \lceil r/2 \rceil$. In the allowed cost $O(M(n) \log n)$, we can assume that $v = h^t$ has been computed. Then we compute $q, w \in F[x]$ with

$$f = qv + w \text{ and } \deg w < \deg v,$$

and then recursively solve "Taylor expansion" for (w, h) (yielding $b_0, \ldots, b_{t-1}$), and for (q, h) (yielding $b_t, \ldots, b_r$). (v) is in Brent & Kung (1978), Lemma 2.1. □

Using fast polynomial arithmetic as quoted above, we can implement the algorithm of Kozen & Landau (1989) very efficiently. For a polynomial $f = x^n + a_{n-1}x^{n-1} + \cdots + a_0 \in F[x]$, we denote by $\tilde{f} = a_0 x^n + \cdots + a_{n-1}x + 1 = x^n \cdot f(1/x)$ the *reversal* of f.

Algorithm **Univariate decomposition.**

Input: $f \in F[x]$ monic of degree $n = rs$, and $r \in \mathbb{N}$, not divisible by $\mathrm{char}(F)$.

Output: The unique normal decomposition (g, h) of f with $\deg g = r$, if such a decomposition exists, and "no decomposition" otherwise.

1. Let $\tilde{f}$ be the reversal of f, and compute $\tilde{h} \in F[x]$ of degree less than s with $\tilde{h}^r \equiv \tilde{f} \bmod x^s$ and $h(0) = 1$. Let $h = x^s \tilde{h}(1/x) \in F[x]$.
2. Compute $b_0, \ldots, b_r \in F[x]$ as in "Taylor expansion".
3. If $b_0, \ldots, b_r \in F$, set $g = \sum_{0 \le i \le r} b_i x^i \in F[x]$ and return (g, h). Otherwise return "no decomposition".

THEOREM 2.2. *Over any field F, algorithm* **Univariate decomposition** *correctly solves* $\mathrm{DEC}^F_{n,r}$, *if* $\mathrm{char}(F)$ *does not divide r. It can be implemented with $O(M(n) \log n)$ arithmetic operations.*

PROOF. Let $f \in F[x]$ be monic of degree $n = rs$. Clearly (g, h) returned by the algorithm is correct, because $f = g \circ h$, h is monic since $\tilde{h}(0) = 1$, and $h(0) = 0$ since $\deg \tilde{h} < s$. So assume that there is a decomposition $f = \hat{g} \circ \hat{h}$, with $\hat{g}, \hat{h}$ monic, and $\hat{h}(0) = 0$. Then f and $\hat{h}^r$ agree on the highest s terms, i.e., $\deg(f - \hat{h}^r) \le n - s$. Let $h_1 = x^s \cdot \hat{h}(1/x)$ be the reversal of $\hat{h}$. Then

$$x^n \hat{h}(1/x)^r = (x^s \hat{h}(1/x))^r = h_1^r,$$

$$\begin{aligned} \deg(f - \hat{h}^r) \le n - s \quad &\Longleftrightarrow \quad x^n \cdot ((f - \hat{h}^r)(1/x)) \equiv 0 \bmod x^s \\ &\Longleftrightarrow \quad \tilde{f} - h_1^r \equiv 0 \bmod x^s. \end{aligned}$$

Since the solution to "Roots" is unique, we have $\tilde{h} = h_1$ and $h = \hat{h}$. Then also $g = \hat{g}$ is computed in step 3.

With the algorithms for "Roots" and "Taylor expansion" from Fact 2.1, the algorithm can be performed with $O(M(n) \log n)$ operations. □

The algorithm shows rationality (as quoted above) and uniqueness of normal univariate decomposition into two polynomials, which will be used in Section 4. The following corollary describes the structure of the set of all decompositions of a fixed polynomial into two decomposition factors, without restriction to normality. Let us call two decompositions $f = g_1 \circ h_1 = g_2 \circ h_2$ *similar* if they differ by an affine linear transformation, i.e., if there exist $c, d \in F$ such that $c \neq 0$, $g_1 = g_2(cx + d)$, and $h_1 = (h_2 - d)/c$.

COROLLARY 2.3. *Let F be a field, $f \in F[x]$ of degree n, and $r \in \mathbb{N}$ with* $\mathrm{char}(F)$ *not dividing r.*

(i) Any two decompositions $f = g_1 \circ h_1 = g_2 \circ h_2$ with $\deg g_1 = \deg g_2 = r$ are similar.

(ii) There exists at most one normal decomposition $f = g \circ h$ with $(f, (g, h)) \in \mathrm{DEC}^F_{n,r}$.

(iii) If $f = g \circ h$ is a decomposition over an extension field K of F, and $h = cx^s + \cdots + d \in K[x]$, where $s = n/r = \deg h$, then $g_1 = g(cx + d)$ and $h_1 = (h - d)/c$ are in $F[x]$, and $f = g_1 \circ h_1$.

PROOF. Let $f = g \circ h$ be any decomposition over K, $a \in F$ the leading coefficient of f, and s, c, d, g_1, h_1 as in (iii). Then

$$a^{-1}f = a^{-1}g_1 \circ h_1$$

is a decomposition, with $a^{-1}f$ and h_1 monic, so that also $a^{-1}g_1$ is monic. Thus this is the unique normal decomposition computed by the algorithm, all polynomials are over F, and also g_1 is over F. This proves all claims. □

Algorithm **Complete decomposition.**

Input: Monic $f \in \mathcal{M} \subset F[x]$ of degree n, not divisible by $\mathrm{char}(F)$.

Output: A complete decomposition of f into indecomposable polynomials.

1. Compute the prime factorization $n = p_1^{e_1} \cdots p_k^{e_k}$ of n. Let $d(n) = (e_1 + 1) \cdots (e_k + 1)$ be the number of divisors of n, and $r_1 = 1 < r_2 < \ldots < r_{d(n)} = n$ the divisors.

2. For $j = 2, \ldots, d(n) - 1$ solve the problem DEC^F_{n,r_j}, with input f. If the first decomposition $(f, (g, h)) \in \mathrm{DEC}^F_{n,r_j}$ is found, apply the algorithm recursively to h, with output $f_2, \cdots, f_k$ such that $h = f_2 \circ f_3 \circ \cdots \circ f_k$. [$f_1 = g$ then is indecomposable.]

3. Return $(f_1, f_2, \ldots, f_k)$.

THEOREM 2.4. *Let $\epsilon > 0$. If* $\mathrm{char}(F)$ *does not divide the degree n of f, algorithm* **Complete decomposition** *computes a complete decomposition of f into indecomposable polynomials with $O(n^{1+\epsilon})$ arithmetic operations.*

PROOF. Clearly the algorithm works correctly. Its cost is $O(d(n) \cdot M(n) \log n)$, since $\deg h$ in the recursive call is a proper divisor of n. We now use that $d(n) = O(n^\epsilon)$ (Hardy & Wright 1962, Theorem 317). □

The algorithm finds the *lexicographically first complete decomposition*, for which the degree sequence $(\deg f_1, \ldots, \deg f_k)$ is lexicographically smallest. Furthermore, each f_i is monic and, except possibly f_1, has constant term zero. The uniqueness in Corollary 2.3 shows that this decomposition exists and is unique.

The number of distinct complete decompositions (with h monic and $h(0) = 0$) is not too large: any polynomial of degree n has at most n distinct decompositions, and if n is the product of the first k prime numbers, then the number $k!$ of decompositions of $f = x^n$ is $\omega(n^{1-\epsilon})$ for every $\epsilon > 0$.

Open Question 2.5. Can one reduce polynomial multiplication to decomposition? Is it possible to improve the running time for decomposition further, say to $O(M(n))$? Given f, g, h, can one compute $g \circ h$, or at least test $f = g \circ h$, deterministically in time $O(M(n))$?

Note that polynomial multiplication is reducible to squaring (if $\text{char}(F) \neq 2$), which is a special case of composition.

Remark 2.6. We have stated Theorem 2.2 only for the case of a field F. The algorithm actually works for an arbitrary commutative ring F with 1, provided that r is a unit in F. Of course, the uniqueness of Corollary 2.3 (ii) may get lost, as in $x^2 \circ (x^2 + ex) = x^2 \circ x^2$, if $\text{char}(F) = 2$ and $e^2 = 0$.

Over the fields of greatest importance in computer algebra, $\mathbb{Q}$ and finite fields, the algorithm can be executed in polynomial time also by Boolean computations, say on a Turing machine or on Booolean circuits. This is trivial over a finite field, where an arithmetic operation can be performed in polynomial time, namely with $O(k \log k \log\log k)$ Boolean operations if $\#F \leq 2^k$. Over $\mathbb{Q}$, we have to show that the binary length of intermediate results is polynomially bounded. Applying naive estimates to iterative methods—such as Kozen & Landau's for the decomposition problem, the Newton iteration used in Fact 2.1 (iii), or the divide-and-conquer for 2.1 (iv)—yields at best "quasi-polynomial" bounds like $n^{\log n}$ times input length. We now prove a polynomial bound.

We represent a rational number a as the quotient of two relatively prime integers $a = b/c$, and call $\max\{\log_2 |b|, \log_2 |c|\}$ the length $l(a)$ of a (assuming $b \neq 0$). Then a can be represented by a string of $O(l(a))$ bits. For a polynomial $f \in \mathbb{Q}[x]$, $l(f)$ is the maximum length of its coefficients. If $a \in \mathbb{Z}$ is the sum of m integers, each of length at most k, then $l(a) \leq \log m + k$.

Theorem 2.7. *Suppose that $f \in \mathbb{Q}[x]$ is monic of degree n and has length $l(f) \leq k$ with $k \geq 2$. Then all rational numbers computed in the algorithm* **Univariate decomposition** *with input f and $r \geq 2$ have length at most $2n^3k$.*

Proof. In the notation of the algorithm, write $\tilde{f} = 1 + f_1$, so that x divides f_1, and $f_1 = a^{-1}f_2 = a^{-1}bf_3$ with $a, b \in \mathbb{N}$, $f_2, f_3 \in \mathbb{Z}[x]$, f_3 primitive, and $\gcd(a, b) = 1$. Then $l(a) \leq nk$ and $l(f_2) \leq nk$. We consider the binomial expansion

$$\tilde{h} \equiv (1 + a^{-1}f_2)^{1/r} \equiv \sum_{0 \leq i} \binom{1/r}{i} a^{-i} f_2^i$$

$$\equiv \frac{1}{r^{s-1}(s-1)!a^{s-1}} \sum_{0 \leq i < s} r^{s-1}(s-1)! \binom{1/r}{i} a^{s-1-i} f_2^i \bmod x^s,$$

where the last sum is in $\mathbb{Z}[x]$. The denominator $u = r^{s-1}\ (s-1)!a^{s-1}$ has length $l(u) \leq (s-1)(nk + \log n)$. For the integer coefficients occurring in the sum, we have

$$\begin{aligned} & l\left((r^{s-1}(s-1)!\binom{1/r}{i}\right) \\ = \;& l\left(r^{s-1-i}\frac{(s-1)!}{i!}\cdot 1\cdot(1-r)\cdot(1-2r)\cdots(1-(i-1)r)\right) \\ \leq \;& (s-1-i)\log r + (s-1-i)\log s + i\log sr = (s-1)\log n. \end{aligned}$$

Recall that $j \in \mathbb{N}$ has $\binom{j-1}{i-1}$ ordered partitions into i positive integers. For any $i < s$, in the binomial expansion of f_2^i each coefficient of an x^j with $j < s$ is the sum of $\binom{j-1}{i-1} < 2^{s-1}$ terms, each of which is the product of at most i coefficients of f_2. This shows that

$$l(a^{s-1-i}f_2^i) \leq (s-1-i)nk + (s-1) + ink = (s-1)(nk+1).$$

Putting this together, we find $u\tilde{h}, uh \in \mathbb{Z}[x]$, and

$$\begin{aligned} l(uh) &= l(u\tilde{h}) \leq (s-1)\log n + (s-1)(nk+1) + \log s \\ &\leq (s-1)(nk + \log n + 2), \\ l(h) &\leq \max\{(s-1)(nk+\log n), (s-1)(nk+\log n + 2)\} \\ &= (s-1)(nk + \log n + 2). \end{aligned}$$

The Taylor expansion problem

$$f = \sum_{0\leq i<r} b_i h^i + h^r,\ b_i \in \mathbb{Q}[x],\ \deg b_i < s$$

is—by equating coefficients of powers of x—equivalent to an $n \times n$-system of linear equations for the rational coefficients of $b_0, \cdots, b_{r-1}$. (We have normalized $b_r = 1$.) Each entry of this system is a coefficient of $f - h^r, h^0, \cdots, h^{r-1}$. Since $uh, uf \in \mathbb{Z}[x]$, multiplication by u^r of each equation yields an equivalent system S with only integral entries. For any $i \leq r$ we have

$$\begin{aligned} l(u^r h^i) &= l(u^{r-i}) + l((uh)^i) \\ &\leq (r-i)(s-1)(nk+\log n) + (n-1) + i(s-1)(nk + \log n + 2) \\ &\leq n(nk + \log n) - 1 \end{aligned}$$

Thus each entry of S has length at most $n(nk + \log n)$. (The -1 covers the coefficients of $f - h^r$.) Cramer's rule and Hadamard's inequality imply that the (unique) solution to S has each component bounded in length by

$$\frac{n}{2}\log n + n\cdot n(nk + \log n) \leq 2n^3 k,$$

so that $l(b_i) \leq 2n^3k$ for any i. □

This theorem easily generalizes to algebraic number fields instead of $\mathbb{Q}$.

3. Very fast parallel decomposition

Kozen & Landau (1986) observe that the general parallelization technique of Valiant *et al.* (1983) applies to their construction, and obtain an arithmetic algorithm of depth $O(\log^2 n)$, in the tame case. This section provides an algorithm of optimal (up to constant factors) depth $O(\log n)$.

For this result, we implement the algorithm of Section 2 fast in parallel. Eberly (1989) shows that division with remainder and "iterated product" of n polynomials of degree at most n can be computed in depth $O(\log n)$ on P-uniform arithmetic Boolean circuits over F. This will be our model for this section; we could also use Eberly's log-space uniform circuits of depth $O(\log n \log\log n)$. The problem is also in Boolean NC over $\mathbb{Q}$ (using bounds like Theorem 2.7) and over finite fields. (Due to the lack of fast parallel Boolean computations for the gcd of integers and of inversion in finite fields, we have to allow the "redundant representation" a/b with $a, b \in \mathbb{Z}$, $b \neq 0$, of field elements, without insisting on $\gcd(a,b) = 1$ over $\mathbb{Q}$ or $b = 1$ in $\mathbb{Z}_p$ (but: $0 \leq a, b < p$).) We start with the problem "Roots" of Fact 2.1 (iii). Note that Newton iteration would only yield depth $O(\log^2 n)$.

LEMMA 3.1. *"Roots" can be solved in depth $O(\log n)$.*

PROOF. By assumption, x divides $f_1 = f - 1$. We have

$$h \equiv (1 + f_1)^{1/r} \equiv \sum_{0 \leq i} \binom{1/r}{i} f_1^i \equiv \sum_{0 \leq i < s} \binom{1/r}{i} f_1^i \bmod x^s.$$

The powers f_1^i can be computed in depth $O(\log n)$. In the model of "non-uniform" arithmetic circuits, the binomial coefficients can be considered as constants in F, and hence given for free. However, they can also be computed in F log-space uniformly, just using the constants 0 and 1 and field operations. This is trivial if $p = \mathrm{char}(F)$ is zero or at least s, by computing the numerator and denominator products separately, and then dividing. If $p < s$, Lemma 3.2 below says that we can replace $1/r$ by $u \in \mathbb{Z}$ in the binomial coefficient, if $ru \equiv 1 \bmod p^{l+1}$, where $l = \lfloor \log_p i \rfloor$. Then we can apply Lucas' (1877) formula:

$$\binom{u}{i} = \binom{u_0}{i_0} \cdots \binom{u_l}{i_l} \bmod p,$$

where $u = u_0 + u_1 p + \cdots + u_l p^l$ with $0 \leq u_j < p$ is the p-ary representation of u, and similarly for i. (The computation of l, u, u_j, i_j takes place in the "Boolean part" of the arithmetic Boolean circuit.) □

LEMMA 3.2. *Let $p \in \mathbb{N}$ be prime, $r, i, m, u, v \in \mathbb{Z}$ with p not dividing r, $i \geq 1$, $l = \lfloor \log_p i \rfloor$, $m > l$, $v \equiv ur \bmod p^m$, and $b = \binom{v/r}{i} \in \mathbb{Q}$. Then b is a p-adic integer (i.e., p does not divide the reduced denominator of b), and $b \equiv \binom{u}{i} \bmod p^{m-l}$ (i.e., the two sides differ by a multiple wp^{m-l} of p^{m-l}, where the reduced denominator of $w \in \mathbb{Q}$ is not divisible by p.)*

PROOF. It is convenient to use the ring

$$\mathbb{Z}_{(p)} = \{s/t \in \mathbb{Q} : s, t \in \mathbb{Z} \text{ and } p \nmid t\}$$

of p-adic integers, with $\mathbb{Z} \subseteq \mathbb{Z}_{(p)} \subseteq \mathbb{Q}$. For this proof, "divisibility" always refers to $\mathbb{Z}_{(p)}$. Choose a bijection

$$\phi : \{1, \ldots, i\} \longrightarrow \{u - i + 1, \ldots, u\}$$

such that

$$\forall j,\ 1 \le j \le i,\ p^l \mid j \Longrightarrow p^l \mid \phi(j).$$

Such a bijection exists. [Choosing some w with $1 \le w \le i$ and $p^l \mid u - i + w$, we can let ϕ map p^l to $u - i + w$, and then the other values preserving the order, except that a wrap-around occurs at u.]

For $1 \le j \le i$, let $\nu(j) = p^{-k}$ if $p^k \mid j$ and $p^{k+1} \nmid j$. Then $k \le l$, $j\nu(j) \in \mathbf{Z}_{(p)}$ is a unit, $(u - v/r)\nu(j) \equiv 0 \bmod p^{m-1}$, and

$$\begin{aligned}
\phi(j)\nu(j) &\equiv (\phi(j) - u + \frac{v}{r})\nu(j) \bmod p^{m-l}, \\
\binom{u}{i} &= \frac{\prod_{1\le j\le i}\phi(j)}{\prod_{1\le j\le i} j} = \frac{\prod_j \phi(j)\nu(j)}{\prod_j j\nu(j)} \\
&\equiv \frac{\prod_j(\phi(j) - u + \frac{v}{r})\nu(j)}{\prod_j j\nu(j)} = \binom{v/r}{i} = b \bmod p^{m-l}. \ \square
\end{aligned}$$

THEOREM 3.3. *Over any field F, the decomposition problem $\mathrm{DEC}^F_{n,r}$, with* char(F) *not dividing r, can be computed on an arithmetic Boolean circuit over F of depth $O(\log n)$.*

PROOF. Using algorithm **Univariate decomposition**, it is now sufficient to solve the "Taylor expansion" problem of Fact 2.1 (iv). If $0 \le i < s$, and

$$f = q_i h^i + r_i \text{ and } \deg r_i < i \deg h$$

is a division with remainder, then $b_i = q_i - hq_{i+1}$. All these computations can be done in depth $O(\log n)$, by Eberly (1989). $\square$

PROPOSITION 3.4. *If* char(F) *does not divide the degree n of f, a complete decomposition of f into indecomposable polynomials can be computed in depth $O(\log n)$.*

PROOF. In the notation of algorithm **Complete decomposition**, solve all problems DEC^F_{n,r_j} in parallel. Suppose that k decompositions $(g_1, h_1), \cdots, (g_k, h_k)$ are found, with degrees $(t_1, s_1), \cdots, (t_k, s_k)$. Order these decompositions so that

$$s_0 = 1 < s_1 < s_2 < \cdots < s_k < s_{k+1} = n.$$

Determine $m \le k$ so that $s_m \le n^{1/2} < s_{m+1}$. If $s_m \ge n^{1/3}$, decompose g_m and h_m recursively. If $s_m < n^{1/3}$ and $s_{m+1} \le n^{2/3}$, decompose g_{m+1} and h_{m+1} recursively. If $s_m < n^{1/3}$ and $s_{m+1} > n^{2/3}$, then compute the decomposition $g_m = e_1 \circ e_2$ with e_2 of minimal degree (at least 2), and decompose e_1 and h_m recursively (using $e_1 = x$ if g_m is indecomposable.) Since $f = e_1 \circ e_2 \circ h_m$, we have $\deg e_1 < n^{1/3}$ in this case; e_2 is indecomposable.

If $D(n)$ denotes the maximal depth for degrees up to n, we have

$$D(n) = O(\log n) + D(n^{2/3}),$$

from which $D(n) = O(\log n)$ follows. $\square$

We show by example that this procedure does not necessarily find the lexicographically first complete decomposition. If $i \ge 1$, $p \in \mathbf{N}$ a prime with $2^i < p \le 2^{2i}$, $n = 2^i p$, and $f = x^n$, then at the top level of the algorithm the first of the three cases occurs with

$$n^{1/3} \le s_m = 2^i \le n^{1/2} < s_{m+1} = p,$$

and the complete decomposition

$$f = g_m \circ h_m = x^p \circ x^{2^i} = x^p \circ x^2 \circ \cdots \circ x^2$$

is computed.

4. Separated polynomials

Let $f_1, f_2 \in F[x]$. Then $f_1(x) - f_2(y) \in F[x, y]$ is called a *separated polynomial.* Finding separated factors of separated polynomials is equivalent to simultaneous decomposition of two polynomials.

FACT 4.1. *(Fried & MacRae 1969b) Let $f_1, f_2, h_1, h_2 \in F[x] \setminus F$ have degrees n_1, n_2, s_1, s_2, respectively. Then $h_1(x) - h_2(y)$ divides $f_1(x) - f_2(y)$ in $F[x, y]$ if and only if there exists $g \in F[x]$ such that $f_i = g \circ h_i$ for $i = 1, 2$. If this is the case, then $\deg g = n_1/s_1 = n_2/s_2 \in$ $\mathbb{N}$.*

A separated factor $h_1(x) - h_2(y)$ is called *normal* if h_1 is monic and $h_1(0) = 0$; any separated factor can be made normal by the affine linear transformation

$$u(x, y)(h_1(x) - h_2(y)) = au(x, y)\left(a^{-1}(h_1(x) - h_1(0)) - a^{-1}(h_2(y) - h_1(0))\right),$$

where a is the leading coefficient of h_1. Let us call $\mathrm{SEP}^F_{n,r}$ the problem of determining, on input $r \in \mathbb{N}$ and two polynomials $f_1, f_2 \in F[x]$ of degree at most n, whether there exists an normal separated factor $h_1(x) - h_2(y)$ of $f_1(x) - f_2(y)$ in $F[x, y]$, with $\deg h_1 = \deg f_1/r$. The literature contains no polynomial-time algorithm to solve this problem, say over $\mathbb{Q}$ or finite fields. One method, requiring exponential time in the worst case, is to compute all irreducible factors of $f_1(x) - f_2(y)$ in $F[x, y]$, and test each product of these factors for being separated. Alagar & Thanh (1985) and Barton & Zippel (1985) based their (exponential-time) decomposition algorithms on this approach, with $f_1 = f_2$. We now turn the fact around and have the following fast algorithm for $\mathrm{SEP}^F_{n,r}$ in the tame case, using univariate decomposition.

Algorithm **Separated factors.**

Input: $r \in \mathbb{N}$ with $\mathrm{char}(F)$ not dividing r, and $f_1, f_2 \in F[x]$ of degrees n_1, n_2, respectively.

Output: All (h_1, h_2) with $h_1(x) - h_2(y)$ a proper normal separated factor of $f_1(x) - f_2(y)$ in $F[x, y]$, and h_1 and $h_2 \in F[x]$ of degrees n_1/r and n_2/r, respectively. If no such polynomials exist: "no separated factors".

1. For $i = 1, 2$ do the following. Let a_i be the leading coefficient of f_i. Compute the normal decomposition $a_i^{-1} f_i = g_i \circ \hat{h}_i$ with $\deg g_i = r$, h_i monic and $\hat{h}_i(0) = 0$. Set $h_1 = \hat{h}_1$. If one of the two polynomials has no such decomposition, return "no separated factors" and stop.

2. Compute the roots $c_1, \cdots, c_t \in F$ of $x^r - a_2/a_1 \in F[x]$ $(0 \le t \le r)$. If $t = 0$, return "no separated factors" and stop.

3. Let $b_1, b_2 \in F$ be the coefficients of x^{r-1} in g_1, g_2, respectively. For $1 \le j \le t$ do steps 4 and 5.

4. Compute $d_j = (b_2 c_j - b_1)/r$.

5. If $a_2 g_2 = a_1 g_1(c_j x + d_j)$, then return h_1 and $h_2 = c_j \hat{h}_2 + d_j$.

6. If for no value of j step 5 was successful, return "no separated factors".

THEOREM 4.2. *Let $n = \max\{n_1, n_2\}$. The algorithm works correctly as described in "Output", and can be performed with $O(r^2 \log^4 r + n \log^2 n \log\log n)$ arithmetic operations in F, plus the computation of all rth roots in F of some $a \in F$.*

PROOF. If h_1 and h_2 are returned in step 5, one checks that $f_i = a_1 g_1 \circ h_i$ for $i = 1, 2$. Thus in view of Fact 4.1, it is sufficient to show:

$$\forall \tilde{g}, \tilde{h}_1, \tilde{h}_2 \in F[x] \; (\deg \tilde{g} = r, f_i = \tilde{g} \circ \tilde{h}_i \text{ for } i = 1,2, \; \tilde{h}_1 \text{ monic with } \tilde{h}_1(0) = 0$$
$$\Longrightarrow \text{ for some value of } j, \; h_1 = \tilde{h}_1 \text{ and } h_2 = \tilde{h}_2 \text{ are returned in step 5.)}$$

So assume that $\tilde{g}, \tilde{h}_1, \tilde{h}_2$ satisfy the hypothesis, let $s = \deg \tilde{h}_2 = n_2/r$, and write $\tilde{h}_2 = \tilde{c}x^s + \cdots + \tilde{d}$. First note that Corollary 2.3(i) and the existence of $\tilde{g}, \tilde{h}_1, \tilde{h}_2$ imply that step 1 successfully computes $g_1, g_2, h_1, \hat{h}_2$. Furthermore, g_1 and g_2 are monic. From

$$g_1 \circ h_1 = a_1^{-1} f_1 = a_1^{-1} \tilde{g} \circ \tilde{h}_1,$$

$$g_2 \circ \hat{h}_2 = a_2^{-1} f_2 = a_2^{-1} \tilde{g} \circ \tilde{h}_2 = a_2^{-1} \tilde{g}(\tilde{c}x + \tilde{d}) \circ (\tilde{h}_2 - \tilde{d})/\tilde{c},$$

and uniqueness of normal decompositions (Corollary 2.3 (ii)) we find

$$a_1^{-1} \tilde{g} = g_1, \quad \tilde{h}_1 = h_1, \quad a_2^{-1} \tilde{g}(\tilde{c}x + \tilde{d}) = g_2, \quad (\tilde{h}_2 - \tilde{d})/\tilde{c} = \hat{h}_2.$$

Comparing coefficients of x^r and x^{r-1} in the third equation, and using the first equation, we obtain

$$a_2^{-1} a_1 \tilde{c}^r = 1, \quad a_2^{-1} \cdot a_1 \tilde{c}^{r-1}(b_1 + r\tilde{d}) = b_2.$$

This shows that $(\tilde{c}, \tilde{d})$ equals (c_j, d_j) for some $j \le t$, and for this value of j, $h_1 = \tilde{h}_1$ and $h_2 = \tilde{c}\hat{h}_2 + \tilde{d} = \tilde{h}_2$ are returned.

The dominating computing costs occur in step 1, with $O(n \log^2 n \log\log n)$ operations in F, and step 5. The equality can be tested by substituting $r+1$ different values from F for x, at a cost of $O(r \log^2 r)$ arithmetic operations (Borodin & Munro 1975, Corollary 4.5.4). If F has at most r elements, we may have to perform this test over a field extension of degree at most $\log(r+1)$ over F, where one arithmetic operation can be done with $O(\log^2 r)$ operations in F. Thus for one value of j, the test has cost $O(r \log^4 r)$, for a total cost of $O(tr \log^4 r) = O(r^2 \log^4 r)$. □

While a separated polynomial of degree n may have $2^n - 2$ proper factors, only few of these are (normal) separated:

COROLLARY 4.3. *Let F be a field, $r \in \mathbb{N}$, $f_1, f_2 \in F[x]$ of degrees n_1, n_2, respectively, $n = \max\{n_1, n_2\}$, $m = \gcd(n_1, n_2)$, $f = f_1(x) - f_2(y) \in F[x, y]$, and $\epsilon > 0$.*

(i) If $\operatorname{char}(F)$ does not divide r, then f has at most r normal separated factors $h_1(x) - h_2(y)$ with h_1 and $h_2 \in F[x]$ of degrees n_1/r and n_2/r, respectively.

(ii) *If* $F = \mathbb{Q}$, *there are at most two normal separated factors of degrees* n_1/r *and* n_2/r, *and at most one if* r *is odd. They can be computed with* $O(n \log^2 n \log\log n)$ *arithmetic operations, plus one root extraction in* $\mathbb{Q}$.

(iii) *If* $\mathrm{char}(F)$ *does not divide* m, *then* f *has* $O(m \log\log m)$ *normal separated factors over* F *(of arbitrary degrees). They can be computed with* $O(n^2 \log^4 n)$ *operations in* F, *plus the extraction of* $O(m \log\log m)$ *roots in* F.

(iv) *If* $F = \mathbb{Q}$, f *has at most* $2d(m) = O(m^\epsilon)$ *normal separated factors, where* $d(m)$ *is the number of factors of* m. *They can be computed with* $O(n^{1+\epsilon})$ *operations in* $\mathbb{Q}$, *plus the extraction of at most* $d(m)$ *roots in* $\mathbb{Q}$.

PROOF. (ii) a_2/a_1 has at most two rth roots in $\mathbb{Q}$. It is sufficient to compute one root c_1; $c_2 = -c_1$ is the other one if r is even.

(iii) We apply **Separated factors** with r running through all divisors of m, and use $\sum_{r|m} r = O(m \log\log m)$ (Hardy & Wright 1962, Theorem 323) and $\sum_{r|m} r^2 = O(m^2)$.

(iv) $d(m) = O(m^\epsilon)$ (Hardy & Wright 1962, Theorem 317). □

The algorithm **Separated factors** requires all rth roots in F of some $a \in F$. Is this really necessary? Let us call ROOT_r^F the problem of finding all rth roots of an input $a \in F$, where we consider r to be the input size.

THEOREM 4.4. *In the tame case, where* $\mathrm{char}(F)$ *does not divide* r, ROOT_r^F *is linear-time reducible to* $\mathrm{SEP}_{r,r}^F$.

PROOF. Let $x^r - a = f_1 \cdots f_m$ be a factorization into monic irreducible factors in $F[x]$, with $f_i = x - c_i$ linear for $1 \le i \le t$, and $f_{t+1}, \cdots, f_m$ nonlinear. For any $f \in F[x]$ of degree n, we denote by

$$\tilde{f} = y^n f(x/y) \in F[x,y]$$

the homogeneous version of f. Then $x^r - ay^r = \tilde{f}_1 \cdots \tilde{f}_m$, and for $c \in F$ we have

$$\begin{aligned} c \text{ is an } r\text{th root of } a &\iff x - c \mid x^r - a \text{ in } F[x] \\ &\iff x - cy \mid x^r - ay^r \text{ in } F[x,y] \\ &\iff x - cy \text{ is a separated factor of } x^r - ay^r. \ \square \end{aligned}$$

Since root-extraction is not a rational process, we now consider Boolean computations over a "computable field"; the most important cases are $\mathbb{Q}$, where roots are easy to compute in polynomial time, or a finite field, where univariate polynomials can be factored in random polynomial time (Berlekamp 1970). On the other hand, Fröhlich & Shepherdson (1955) exhibit computable fields of characteristic zero over which the existence of (square) roots of a given field element is undecidable.

COROLLARY 4.5. (i) *There exist fields* F *over which* $\mathrm{SEP}_{n,r}^F$ *is uncomputable.*

(ii) $\mathrm{SEP}_{n,r}^{\mathbb{Q}}$ *can be computed in polynomial time.*

(iii) *If* F *is a finite field with* q *elements, then* $\mathrm{SEP}_{n,r}^F$ *can be computed by a probabilistic algorithm using time polynomial in* $\log q$ *and* n.

PROOF. For (ii), use Theorem 2.7. □

EXAMPLE 4.6. It is a bit surprising that although decomposition is rational (Corollary 2.3) and separated factorization is equivalent to simultaneous decomposition of two univariate polynomials (Fact 4.1), separated factorization is **not** rational. The algorithm indicates how to generate examples such as the following. The separated polynomial $f = x^2 + y^2 + 10y + 25 \in \mathbb{Q}[x, y]$ is irreducible over $\mathbb{Q}$, but has the separated factorization $f = (x - iy - 5i) \cdot (x + iy + 5i)$ over $\mathbb{Q}[i] \subset \mathbb{C}$, where $i = \sqrt{-1}$. The corresponding simultaneous decomposition is $f_1 = x^2 = x^2 \circ x$, $f_2 = -x^2 - 10x - 25 = x^2 \circ (ix + 5i)$. In step 2 of the algorithm, we would obtain the normal decomposition $a_2^{-1} f_2 = x^2 + 10x + 25 = (x^2 + 10x + 25) \circ x$.

In general, we have the following description of all separated factorizations, rational or not. Let F be a field, and K an algebraic closure of F. Corollary 4.3 (i) and (iii), with F replaced by K, give a bound on the number of factorizations.

COROLLARY 4.7. *Let F, K be as above, $r \in \mathbb{N}$, $f_1, f_2 \in F[x]$ of degrees n_1, n_2, respectively, $n = \max\{n_1, n_2\}$, a_i the leading coefficient of f_i, for $i = 1, 2$, and $a = a_2/a_1$.*

(i) If h_1 and $h_2 \in K[x]$ have degrees n_1/r and n_2/r, respectively, and $h_1(x) - h_2(y)$ is a separated factor of $f_1(x) - f_2(y)$, then $h_1, h_2 \in F[b][x]$ for some $b \in K$ with $b^r = a$.

(ii) A representative set of all h_1, h_2 as in (i), up to conjugation over F, can be computed with $O(M(r)(r^2 \log^4 r + n \log^2 n \log\log n))$ operations in F, plus the factorization of $x^r - a$ in $F[x]$.

PROOF. (i) follows from (the proof of) Theorem 4.2. For (ii), let $x^r - a = g_1 \cdots g_m$ be a factorization into monic irreducible polynomials in $F[x]$, and $d_i = \deg g_i$. ($g_1, \ldots, g_m$ are pairwise distinct.) If $\alpha_i = x \bmod g_i$ is an rth root of a in $F_i = F[x]/(g_i)$, for any i, then any (h_1, h_2) as in (i) is found by Algorithm **Separated factors** with some α_i substituted for c_i in steps 4 and 5 (up to conjugates over F). One arithmetic operation in F_i can be simulated by $O(M(d_i))$ operations in F. Since $\sum_{1 \le i \le m} M(d_i) \le M(r)$ (assuming that $M(d_i)/d_i \le M(r)/r$, which is the case for the choice made in Section 2), the claim for the total cost follows. □

If $\varsigma \in K$ is a primitive rth root of unity, $b \in K$ with $b^r = a$, and $L = F[\varsigma, b] \subseteq K$, then $[L : F] \le r\phi(r)$, and all h_1 and h_2 as above are in $L[x]$. One operation in L can be simulated with $O(M(r\phi(r)))$ operations in F, and the cost of performing the algorithm for Corollary 4.7 in L would be as in (ii) above, except that $M(r)$ is replaced by $M(r\phi(r))$.

5. Multivariate polynomials

There are several generalizations of the decomposition problem to multivariate polynomials. We solve the following type of problem:

$$f = g \circ h, \; f, h \in F[x_1, \ldots, x_m], \; g \in F[x]$$

by a simple linearly converging Newton method: substitute for $x_2, \cdots, x_m$, solve the univariate problem, and lift the unique solution.

We first have to make some normalizations. For $\alpha \in F$ we have

$$g \circ h = [g \circ (x - \alpha)] \circ [(x + \alpha) \circ h],$$

so that we may assume throughout this section that $h(0,\ldots,0)=0$. We define the set $\mathcal{M}_m$ of polynomials which are *strongly monic in* x_1 as follows:

$$\begin{aligned}\mathcal{M}_m &= \{f=\sum_{0\le i\le n} f_i x_1^i \in F[x_1,\ldots,x_m]: \\ &\quad n\in \mathbb{N}, f_0,\cdots,f_n\in F[x_2,\ldots,x_m],\ f_n=1,\ \deg f=n\},\end{aligned}$$

where $\deg f$ is the total degree of f. Thus $\mathcal{M}=\mathcal{M}_1$ (identifying x with x_1). The *normal decomposition problem* is

$$\begin{aligned}\mathrm{DEC}^F_{n,r,m} &= \{(f,(g,h))\in \mathcal{M}_m\times(\mathcal{M}\times\mathcal{M}_m): \\ &\quad f=g\circ h,\ \deg f=n,\ \deg g=r, \text{ and } h(0,\ldots,0)=0\}.\end{aligned}$$

Thus $\mathrm{DEC}^F_{n,r,1}=\mathrm{DEC}^F_{n,r}$. If $(f,(g,h))\in\mathrm{DEC}^F_{n,r,m}$, then $f=g\circ h$ (or (g,h)) is called a *normal decomposition* of f. Corollary 5.3 (iv) below states that any f has at most one normal decomposition in the tame case.

We consider the ideal $\mathbf{x}\subseteq F[x_1,\ldots,x_m]$ generated by $x_2,\cdots,x_m$. An $\mathbf{x}$-homogeneous polynomial of degree d is of the form

$$\sum_{j_2+\cdots+j_m=d} a_j x_2^{j_2}\cdots x_m^{j_m},$$

with $a_j\in F[x_1]$; 0 is homogeneous of any degree. Every nonzero $f\in F[x_1,\ldots,x_m]$ has a unique $\mathbf{x}$-homogeneous representation $f=\sum_{i\le\deg f} f_i$ with f_i $\mathbf{x}$-homogeneous of degree i. We write $f_i=H_i(f)$.

The following algorithm solves the normal decomposition problem in the tame case.

Algorithm **Multivariate normal decomposition.**
Input: $f\in\mathcal{M}_m\subset F[x_1,\ldots,x_m]$ of degree $n=rs$, and $r\in\mathbb{N}$, not divisible by $\mathrm{char}(F)$.

Output: The unique normal decomposition (g,h) of f with $\deg g=r$, if such a decomposition exists, and "no decomposition" otherwise.

1. Compute $f_0=f(x_1,0,\cdots,0)$, and $g\in F[x]$, $h_0\in F[x_1]$ with $(f_0,(g,h_0))\in\mathrm{DEC}^F_{n,r,1}$ by algorithm **Univariate decomposition.** If no such g,h_0 exist, return "no decomposition" and stop.

2. Set $s=n/r\in\mathbb{N}$, $k_0=h_0$, and $t=(\partial g/\partial x)\circ h_0\in F[x_1]$. [We will see that $t\neq 0$.]

3. For i from 1 to s perform step 4. [This Newton step determines $h_i=H_i(h)$ and $k_i=\sum_{j\le i}h_j$.]

4. Compute $u_i=H_i(f-g\circ k_{i-1})$. If t does not divide u_i (i.e., each coefficient at a monomial in $x_2,\cdots,x_m$), then return "no decomposition" and stop. Otherwise set $h_i=u_i/t$ and $k_i=k_{i-1}+h_i$.

5. Set $h=k_s$, and return (g,h) if $f=g\circ h$, and "no decomposition" otherwise.

THEOREM 5.1. *In the tame case, every $f\in\mathcal{M}_m\subset F[x_1,\ldots,x_m]$ has at most one normal decomposition. The algorithm correctly decides existence, and computes the normal decomposition, if it exists.*

PROOF. By assumption, p does not divide r, so that $g' = \partial g/\partial x \neq 0$, and since h_0 is a nonconstant polynomial, also $t = g' \circ h_0 \neq 0$ in step 2.

Let $f = \tilde{g} \circ \tilde{h}$ be any normal decomposition, and $f = \sum f_j$ and $\tilde{h} = \sum \tilde{h}_j$ the $\mathbf{x}$-homogeneous representations. It is sufficient to show that the algorithm outputs (g, h) with $g = \tilde{g}$ and $h_i = \tilde{h}_i$, by induction on i.

For $i = 0$, we have $(f_0, (g, \tilde{h}_0)) \in \mathrm{DEC}^F_{n,r,1}$, and Corollary 2.3 (ii) implies that $g = \tilde{g}$ and $h_0 = \tilde{h}_0$ are computed in step 1. Now let $i \geq 1$. Taylor expansion of g around an indeterminate y implies that there exists $G \in F[x, y]$ such that

$$g(x) = g(y) + g'(y) \cdot (x - y) + G \cdot (x - y)^2$$

holds in $F[x, y]$. Substituting $k_{i-1} = \sum_{j<i} \tilde{h}_j$ for y (where we have used the induction hypothesis) and $y + \tilde{h}_i$ for x we find

$$g \circ \tilde{h} \equiv g \circ \sum_{j \leq i} \tilde{h}_j \equiv g \circ k_{i-1} + (g' \circ k_{i-1}) \cdot \tilde{h}_i \bmod \mathbf{x}^{i+1},$$

since $(x - y)^2 = \tilde{h}_i^2 \equiv 0 \bmod \mathbf{x}^{i+1}$. Now

$$g' \circ k_{i-1} = g' \circ \sum_{j<i} h_j \equiv g' \circ h_0 = t \bmod \mathbf{x},$$

since $h_j \equiv 0 \bmod \mathbf{x}$ for $j \geq 1$. So we have

$$0 = f - g \circ \tilde{h} \equiv \sum_{j \leq i} f_j - (g \circ k_{i-1} + t\tilde{h}_i) \bmod \mathbf{x}^{i+1},$$

$$u_i = f_i - H_i(g \circ k_{i-1}) = t\tilde{h}_i.$$

This shows that u_i is divisible by t, and indeed $h_i = \tilde{h}_i$. □

For an arbitrary polynomial $f \in F[x_1, \ldots, x_m]$ we use substitutions σ of the form $\sigma f = f(x_1, x_2 + \sigma_2 x_1, \cdots, x_m + \sigma_m x_1)$, with $\sigma = (\sigma_2, \cdots, \sigma_m) \in F^{m-1}$ to make f strongly monic in x_1. If $h(0, \ldots, 0) = 0$, then also $(\sigma h)(0, \ldots, 0) = 0$. The substitution $\sigma^{-1} = (-\sigma_2, \cdots, -\sigma_m)$ is inverse to σ, with $\sigma^{-1}\sigma f = f$ for any $f \in F[x_1, \ldots, x_m]$.

Algorithm **Multivariate decomposition.**

Input: $f \in F[x_1, \ldots, x_m]$ of (total) degree $n = rs$, and $r \in \mathbb{N}$, not divisible by $\mathrm{char}(F)$.

Output: $g \in F[x]$ and $h \in F[x_1, \ldots, x_m]$ with $f = g \circ h$ and $\deg g = r$, if such a decomposition exists, and "no decomposition" otherwise.

1. Choose a substitution $\sigma \in F^{m-1}$ such that $\deg_{x_1} \sigma f = n$.

2. Let $a \in F$ be the leading coefficient of σf with respect to x_1, and $\tilde{f} = a^{-1}\sigma f$. [Then $\tilde{f} \in \mathcal{M}_m$.]

3. Call Algorithm **Multivariate normal decomposition** with input $\tilde{f}$. If no decomposition of $\tilde{f}$ exists, return "no decomposition". If $\tilde{f} = \tilde{g} \circ \tilde{h}$ is returned, return $g = a\tilde{g}$ and $h = \sigma^{-1}\tilde{h}$, where $\sigma^{-1} = (-\sigma_2, \cdots, -\sigma_m)$ is the substitution inverse to σ.

THEOREM 5.2. *Assume that a substitution σ is chosen as required in step 1 of the algorithm. Then the algorithm correctly determines whether $f \in F[x_1, \ldots, x_m]$ has a decomposition with the required degrees, and if so, computes a decomposition.*

PROOF. Clearly any answer (g, h) of **Multivariate decomposition** is indeed a decomposition $f = g \circ h$ with the required degrees. Now suppose that $f \in F[x_1, \ldots, x_m]$ has a decomposition $f = \hat{g} \circ \hat{h}$, with the required degrees, and $\hat{h}(0, \ldots, 0) = 0$. Let c be the leading coefficient of $\sigma\hat{h}$ with respect to x_1. Since $\sigma f = \hat{g} \circ \sigma\hat{h}$ has degree n in x_1, c is in F. $\tilde{f}$ is in $\mathcal{M}_m$, and

$$\tilde{f} = a^{-1}\sigma f = a^{-1}\hat{g} \circ \sigma\hat{h} = a^{-1}\hat{g}(cx) \circ c^{-1}\sigma\hat{h}$$

is a normal decomposition, with $\tilde{f}$ and $c^{-1}\sigma\hat{h}$ monic in x_1, and $c^{-1}\sigma\hat{h}(0, \ldots, 0) = 0$. Thus also $a^{-1}\hat{g}(cx)$ is monic, and the unique solutions $\tilde{g} = a^{-1}\hat{g}(cx)$ and $\tilde{h} = c^{-1}\sigma\hat{h}$ are returned by **Multivariate normal decomposition**. Thus the algorithm returns the correct decomposition $f = g \circ h = \hat{g}(cx) \circ c^{-1}\hat{h}$. □

A "lucky substitution" σ in step 1 is easy to find:

LEMMA 5.3. *Let $f \in F[x_1, \ldots, x_m]$ have degree n, $A \subseteq F$ finite, and a substitution σ uniformly chosen at random in A^{m-1}. Then $\deg_{x_1} \sigma f = n$ with probability at least $1 - n/\#A$.*

PROOF. For $0 \leq i \leq n$, let $u_i \in F[x_2, \cdots, x_m]$ be the homogeneous part of (highest) degree $n - i$ of the coefficient of x_1^i in f. Thus the homogeneous part of f of (total) degree n is $\sum u_i x_1^i \neq 0$, and by the homogeneity, also $u = \sum u_i \in F[x_2, \cdots, x_m]$ is nonzero, and of degree at most n. Now we have for $\sigma \in F^{m-1}$

$$\begin{aligned} &\deg_{x_1} \sigma f = n \\ &\iff \deg f(x_1, \sigma_2 x_1, \cdots, \sigma_m x_1) = \deg[(\sigma f)(x_1, 0, \cdots, 0)] = n \\ &\iff u(\sigma_2, \cdots, \sigma_m) \neq 0. \end{aligned}$$

The claim now follows from Schwartz (1980). □

Over small fields, no lucky substitutions may exist. For example, $f = x_1^2 x_2 + x_1 x_2^2 \in \mathbf{Z}_2[x_1, x_2]$ has no substitution $\sigma \in \mathbf{Z}_2$ with $\deg_{x_1} \sigma f = 3$.

We now describe the structure of the set of multivariate decomposition of a fixed polynomial, and show rationality as in the univariate case. Let $n = rs \in \mathbf{N}$. Consider two decompositions $f = g_1 \circ h_1 = g_2 \circ h_2$, with $g_1, g_2 \in F[x]$, $f, h_1, h_2 \in F[x_1, \ldots, x_m]$, $\deg f = n > r = \deg g_1 = \deg g_2 \geq 2$, and, as always, $h_1(0, \cdots, 0) = h_2(0, \cdots, 0) = 0$. Let us call the two decompositions *similar* if there exists $c \in F \setminus \{0\}$ such that $g_2 = g_1(cx)$ and $h_2 = c^{-1}h_1$. The constant of similarity c is uniquely determined.

COROLLARY 5.4. *In the tame case, where* char(F) *does not divide r, the following hold.*

(i) *Any two multivariate decompositions are similar.*

(ii) *If the first decomposition factors g_1, g_2 are monic, then the constant of similarity is an rth root of unity.*

(iii) *If $c \in F$ is an rth root of unity and $f = g \circ h$ a decomposition with g monic, then $f = g(cx) \circ c^{-1}h$ is a decomposition of the same form.*

(iv) Any $f \in \mathcal{M}_m$ has at most one normal decomposition $f = g \circ h$ with $h \in \mathcal{M}_m$.

(v) If $f = g \circ h$ is a decomposition over some extension field K of F, $h(0, \cdots, 0) = 0$, and $c \in K$ a nonzero coefficient of h, then $g(cx)$ and $c^{-1}h$ have coefficients in F.

PROOF. (i) Suppose that we have two decompositions $f = g_1 \circ h_1 = g_2 \circ h_2$, and first assume that $\#F > n = \deg f$. By Lemma 5.3 there exists a substitution $\sigma \in F^{m-1}$ such that $\deg_{x_1} \sigma f = n$. The proof of Theorem 5.2 shows that both decompositions are similar to the output of **Multivariate decomposition**, and thus similar to each other. In the general case, consider a field extension $K \supset F$ with $\#K > n$. The above implies that $g_2 = g_1(cx)$ and $h_2 = c^{-1}h_1$ for some $c \in K \setminus \{0\}$. Comparing coefficients in h_1 and h_2 of some monomial that occurs with nonzero coefficients (both in F) shows that $c \in F$.

(v) Suppose that $f = g \circ h$ is a decomposition over some $K \supset F$, with $h(0, \cdots, 0) = 0$. First assume $\#F > n$, and let $\sigma \in F^{m-1}$ be chosen in step 1 of the algorithm with $\deg_{x_1} \sigma f = n$. Let $a \in F \setminus \{0\}$ and $b \in K \setminus \{0\}$ be the leading coefficients with respect to x_1 of σf and σh, respectively. Then

$$\frac{\sigma f}{a}(x, 0, \cdots, 0) = \frac{g}{a} \circ \sigma h(x, 0, \cdots, 0) = \frac{g}{a}(bx) \circ b^{-1} \sigma h(x, 0, \cdots, 0),$$

and $g(bx)$ and $b^{-1}\sigma h(x, 0, \cdots, 0)$ have coefficients in F, by Corollary 2.3 (iii). This univariate decomposition is computed in step 2 of the algorithm. As proven in Theorem 5.2 (over K), the algorithm then computes the decomposition $f = g(bx) \circ b^{-1}h$. However, all steps of the algorithm are rational, so that indeed $g(bx)$ and $b^{-1}h$ are over F. Then also $g(cx) = g(bx) \circ \frac{c}{b}x$ and $c^{-1}h = \frac{b}{c}b^{-1}h$ are over F for any nonzero coefficient c of h.

If $\#F \leq n$, we choose two finite algebraic extension fields $L_1, L_2 \subseteq L$ of F with $\#L_1, \#L_2 > n$ and $[L_1 : F], [L_2 : F]$ two distinct primes, where L is an algebraic closure of F. We may also assume $K \subseteq L$. For $i = 1, 2$, we apply the above argument with L_i for F and the join of K and L_i for K. It follows that $g(cx)$ and $c^{-1}h$ are over $L_1 \cap L_2 = F$. □

We now want to evaluate the cost of the algorithm in three data structures for multivariate polynomials: the dense, sparse, and circuit representations (also called the straight-line representation). These representations are discussed in von zur Gathen (1985).

THEOREM 5.5. *In the tame case, multivariate polynomials can be decomposed in time polynomial in the length of the dense representation, and randomly in time polynomial in the length of a circuit representation.*

PROOF. Let $A \subseteq F$ have $n + 1$ elements. In the dense representation, one can choose $\sigma_2, \cdots, \sigma_m$ deterministically one after the other, making sure that $u(\sigma_2, \cdots, \sigma_i, x_{i+1}, \cdots, x_m)$ is nonzero, with $u \in F[x_2, \cdots, x_m]$ as in the proof of Lemma 5.3. The other steps of the algorithm can clearly be performed in a polynomial number of arithmetic operations.

For the circuit representation, we use $A \subseteq F$ with $2n$ elements, choose $(\sigma_2, \cdots, \sigma_m) \in A^{m-1}$ randomly, compute the coefficients of $f_0 = f(x_1, \sigma_2 x_1, \cdots, \sigma_m x_1)$, and test "$\deg f_0 = n$". By Section 2 and with Kaltofen's (1986) general techniques for manipulating arithmetic circuits, this and the other steps can be performed in random polynomial time.

If F has less than $2n$ elements, one has to perform the algorithm in a field extension K of F of degree at most $\log 2n$. Then scaling any decomposition by a nonzero coefficient yields a decomposition over F (Corollary 5.4 (v)). □

Moenck (1976) shows how to multiply two polynomials of degree at most d_1 and d_2, respectively, in each of m variables with $O(m(d_1+d_2+1)^m \log d)$ operations, assuming that the field supports a Fourier Transform. With this routine, the algorithm can be performed with $O(mn(n+1)^m \log n)$ operations, which is not much more than linear in the corresponding input size $(n+1)^m$. This compares favorably to the estimate of Dickerson's (1987) algorithm, which uses less than N^3 operations if the dense representation of f has N terms. (A variable-by-variable lifting would be appropriate for this input representation, where the degree in each variable is bounded.)

Unfortunately, for the sparse representation—the most intuitive one—it is conceivable that the obvious implementation of the algorithm uses more than polynomial time. This might be the case with $f, h \in F[x_1, \ldots, x_m]$ containing a small number t of nonzero monomials, $g = \sum g_i x^i \in F[x]$ such that $f = g \circ h$, and some h^i with $g_i \neq 0$ having more than polynomial in t many nonzero monomials. Another case might be an indecomposable f with few nonzero terms for which a "dense" h is computed. However, the algorithm might work reasonably well in practice also in the sparse representation, hoping that such bad examples do not occur too often (at least for decomposable f).

OPEN QUESTION 5.6. *Do such examples exist?*

6. Conclusion

We have exhibited fast polynomial decomposition algorithms for certain univariate and multivariate problems. Besides the open questions mentioned in the text (and the more difficult "wild case"), a next goal would be to elucidate the structure of (and find algorithms for) rational decompositions $f = g \circ h$ with $f, g, h \in F(x)$, and different multivariate polyhomial decompositions, such as $f = g(h_1, h_2)$ with $f, h_1, h_2 \in F[x]$.

References

Alagar, V. S., Thanh, M. (1985). Fast polynomial decomposition algorithms. In *Proc. EUROCAL 85, Lecture Notes in Computer Science* **204**, 150-153.

Barton, D. R., Zippel, R. (1985). Polynomial decomposition algorithms. *J. Symb. Comp.* **1**, 159-168.

Berlekamp, E. R. (1970). Factoring polynomials over large finite fields. *Math. Comp.* **24**, 713-735.

Borodin, A., Munro, I. (1975). *The computational complexity of algebraic and numeric problems.* American Elsevier: New York.

Brent, R. P., Kung, H. T. (1978). Fast algorithms for manipulating formal power series. *J. Assoc. Comput. Mach.* **25**, 581-595.

Cade, J. J. (1985). A new public-key cipher which allows signatures. *Proc. 2nd SIAM Conf. on Appl. Linear Algebra*, Raleigh NC.

Cantor, D. G., Kaltofen, E. (1987). Fast multiplication of polynomials over arbitrary rings. Tech. Rep. 87-35, Dept. of Computer Science, Rensselaer Polytechnic Institute, 16 pp. *Acta Inform.*, in press.

Dickerson, M. (1987). Polynomial decomposition algorithms for multivariate polynomials. Tech. Rep. 87-826, Department of Computer Science, Cornell University, Ithaca NY.

Dorey, F., Whaples, G. (1974). Prime and composite polynomials. *J. Algebra* **28**, 88-101.

Eberly, W. (1989). Very fast parallel matrix and polynomial arithmetic. *SIAM J. Comput.* **18**, 955-976.

Engstrom, H. T. (1941). Polynomial substitutions. *Amer. J. Math* **63**, 249-255.

Fried, M. D., MacRae, R. E. (1969a). On the invariance of chains of fields. *Ill. J. Math.* **13**, 165-171.

Fried, M. D., MacRae, R. E. (1969b). On curves with separated variables. *Math. Ann.* **180**, 220-226.

Fröhlich, A., Shepherdson, J. C. (1955). Effective Procedures in Field Theory. *Phil. Trans. Royal Soc., Ser. A* **248**, 407-432.

von zur Gathen, J. (1984). Parallel algorithms for algebraic problems. *SIAM J. Comput.* **13**, 802-824.

von zur Gathen, J. (1985). Irreducibility of multivariate polynomials. *J. Computer System Sciences* **31**, 225-264.

von zur Gathen, J. (1986). Parallel arithmetic computations: a survey. *Proc. 12th Int. Symp. Math. Foundations of Computer Science*, Bratislava, *Springer Lecture Notes in Computer Science* **233**, pp. 93-112.

von zur Gathen, J. (1988). Functional decomposition of polynomials: the wild case. Manuscript.

von zur Gathen, J., Kozen, D., Landau, S. (1987). Functional decomposition of polynomials. *Proc. 28th Ann. IEEE Symp. Foundations of Computer Science*, Los Angeles CA, pp. 127-131.

Gutiérrez, J., Recio, T., Ruíz de Velasco, C. (1989). Polynomial decomposition algorithm of almost quadratic complexity. *Springer Lecture Notes in Computer Science* **357**, 471-476.

Hardy, G. H., Wright, E. M. (1962). *An introduction to the theory of numbers.* Clarendon Press: Oxford.

Hasse, H. (1980). *Number Theory.* Grundlehren der math. Wiss. **229**, Springer Verlag.

Kaltofen, E. (1988). Greatest Common Divisors of Polynomials Given by Straight-line Programs. *J. Assoc. Comput. Mach.* **35**, 231-264.

Kozen, D., Landau, S. (1989). Polynomial decomposition algorithms. Tech. Rep. 86-773, Department of Computer Science, Cornell University, Ithaca NY. *J. Symb. Comp.* **7** (1989), 445-456.

Levi, H. (1942). Composite polynomials with coefficients in an arbitrary field of characteristic zero. *Amer. J. Math.* **64**, 389-400.

Lucas, E. (1877). Sur les congruences des nombres eulériens et des coefficients différentiels des fonctions trigonométriques, suivant un module premier. *Bull. Soc. Math. France* **6**, 49-54.

Moenck, R. T. (1976). Practical fast polynomial multiplication. *Proc. ACM Symp. Symbolic and Algebraic Computation*, pp. 136-148.

Ritt, J. F. (1922). Prime and composite polynomials. *Trans. Amer. Math. Soc.* **23**, 51-66.

Ritzmann, R. (1984). Ein numerischer Algorithmus zur Komposition von Potenzreihen und Komplexitätsschranken für die Nullstellenberechnung von Polynomen. Inaugural-Dissertation, Universität Zürich.

Schönhage, A. (1977). Schnelle Multiplikation von Polynomen über Körpern der Charakteristik 2. *Acta Informatica* **7**, 395-398.

Schwartz, J. T. (1980). Fast probabilistic algorithms for verification of polynomial identities. *J. Assoc. Comput. Mach.* **27**, 701-717.

Valiant, L., Skyum, S., Berkowitz, S., Rackoff, C. (1983). Fast parallel computation of polynomials using few processors. *SIAM J. Comput.* **12**, 641-644.

Computing with Polynomials Given By Black Boxes for Their Evaluations: Greatest Common Divisors, Factorization, Separation of Numerators and Denominators*

ERICH KALTOFEN

Rensselaer Polytechnic Institute, Department of Computer Science, Troy, New York 12180-3590; `kaltofen@cs.rpi.edu`

BARRY M. TRAGER

I.B.M., T. J. Watson Research Center, Yorktown Heights, New York 10598; `bmt@ibm.com`

(*Received* 17 *February* 1989)

Algorithms are developed that adopt a novel implicit representation for multivariate polynomials and rational functions with rational coefficients, that of black boxes for their evaluation. We show that within this representation the polynomial greatest common divisor and factorization problems, as well as the problem of extracting the numerator and denominator of a rational function, can all be solved in random polynomial-time. Since we can convert black boxes efficiently to sparse format, problems with sparse solutions, e.g., sparse polynomial factorization and sparse multivariate rational function interpolation, are also in random polynomial time. Moreover, the black box representation is one of the most space efficient implicit representations that we know. Therefore, the output programs can be easily distributed over a network of processors for further manipulation, such as sparse interpolation.

1. Introduction

We introduce algorithms that manipulate multivariate polynomials and rational functions that are given by "black boxes": a black box is an object which takes as input a value

*This material is based on work supported by the National Science Foundation under Grants No. CCR-87-05363 and No. CDA-88-05910, and by an IBM faculty development award (first author). A preliminary version of this paper appears in the *Proceedings of the 29th Symposium on Foundations of Computer Science*, IEEE, pp. 296–305 (1988).

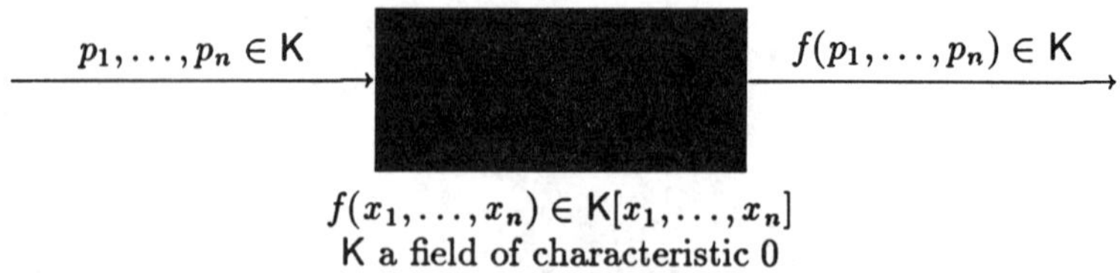

Figure 1: Black box representation of a polynomial.

for each variable, and then produces the value of the polynomial or rational function it represents at the specified point (see Figure 1).

The algorithms' outputs are procedures which will evaluate the answer polynomials at arbitrary points (supplied as the input). These procedures make oracle calls to the black boxes given as the input to the algorithm to evaluate them at certain points dependent on the inputs to the procedure. One result is that we can construct an evaluation procedure for the greatest common divisor of a set of polynomials given their black box representations. A main result is the construction of an evaluation program for all irreducible factors of a black box polynomial (see Figure 2).

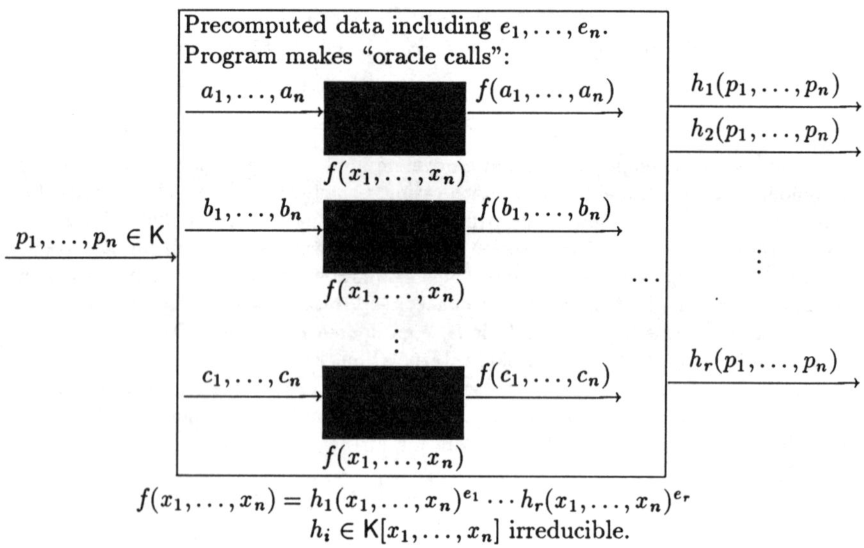

Figure 2: The program for evaluating the irreducible factors of a black box polynomial.

Finally, from a black box for a multivariate rational function we construct an evaluation box for both its reduced numerator and denominator. All three algorithms normalize the constructed polynomials in a deterministic fashion, that is querying the resulting boxes repeatedly will evaluate one and the same associate (scalar multiple) of the goal polynomial. Moreover, for rational coefficients the constructions of the new evaluation boxes are accomplished in polynomial-time with respect to total degree and coefficient size of the given polynomials. The algorithms presented are Monte-Carlo: that is their output—the constructed evaluation programs—procedures are correct with controllably high probability.

Previous results on manipulating implicitly represented multivariate polynomials and rational functions consider straight-line programs as the representation model (Baur and Strassen 1983), (Valiant 1982), (von zur Gathen 1985), (Kaltofen 1986b, 1987, and 1989). The polynomial factorization result for straight-line programs, for instance, performs transformations on each individual assignment in the input program, and cannot be generalized to the black box representation in a straightforward fashion. We employ a new idea for this problem by introducing an extra variable for two different purposes. When setting that variable to 0, the factors map to precomputed fixed univariate factors. We thus can not only identify the bivariate factors independently of the evaluation that is requested, but also normalize them with respect to scalar multiplication. The other purpose of this second variable is that by setting it to 1 we obtain the values of the factors at a given point.

It is, of course, crucial that the program for the irreducible factors evaluates a fixed associate for each multivariate factor. Moreover, the program is with controllably high probability correct, that is it then will always return the correct evaluations of the factors, independently of the input values an adversary may have chosen for the variables. We also impose this requirement on the construction of a program for the numerator and denominator of a rational function.

The solution for numerator and denominator separation found for the straight-line model (Kaltofen 1988) requires stepping through each instruction. Our innovation first uses a mixed radix Padé approximation (see, for instance (von zur Gathen 1986)), that replaces the Taylor series approximation by interpolation. Second, it again makes use of the idea of a multipurpose second variable. However, for that problem the construction of the evaluation procedure for numerators and denominators is substantially more complex. The reason is that one has to deal with zero points of the denominator. Our assumption is that the black box for the rational function indicates in its output if one tries to evaluate at a root of the denominator, and produces a value at all other points. We remark that if the program for evaluating the numerator and denominator is constructed correctly by our algorithm, that with high probability, it will find the values of both the numerator and the denominator polynomials at all values for the variables, even those that are zeros for the denominator.

Our constructed programs with oracle calls to the input black boxes are in many ways superior to the straight-line program model. From a theoretical point of view, these programs can be rapidly converted to sparse format using any of the new sparse polynomial interpolation algorithms (Ben-Or and Tiwari 1988), (Zippel 1990), (Kaltofen and Lakshman 1988), and (Grigoryev et al. 1988). Our black box factorizer thus has, for example, resurrected interpolation as a means of factoring polynomials. That method was last mentioned in the first edition of Knuth's book (1981), but gave way to Hensel lifting for reason of exponential combinatorial blow-up in the identification problem of the factor images.

Having overcome this problem here, we can use sparse interpolation to an added advantage, namely we can distribute the interpolation process over a series of processors using Ben-Or's and Tiwari's (1988) algorithm. Since the polynomials are given by black boxes, the amount of data that needs to be sent to the individual processors is very small, which makes such a distributed computation scheme quite practical. From a theoretical point of view, we also obtain a randomized $\mathcal{NC}$ reduction (Cook 1985) for computing the sparse factors of multivariate polynomials to the problem of factoring univariate polynomials over the coefficient field, resolving an open problem in (von zur Gathen and Kaltofen 1985). Over the rational numbers we get, for example, a randomized $\mathcal{NC}$ algorithm for computing all sparse factors of a multivariate polynomial that have fixed degree.

We can also apply sparse interpolation to the program that we obtain for the numerator and denominator of a black box rational function. Furthermore, if we are given correct upper bounds for the number of non-zero monomials in the numerator and denominator, then by a deterministic zero test (Zippel 1990) we can verify the computed sparse numerator and denominator. Therefore we have a Las Vegas randomized polynomial-time solution to the sparse rational interpolation problem, that is one that always returns the correct answer and has expected polynomial-time complexity.

Before discussing our results, let us mention a linear algebra setting in which black box representation also can be successfully applied. We are given the coefficient matrix A of a linear system in terms of a black box that will multiply this matrix with a chosen vector y (see Figure 3). For simplicity, assume that $A \in \mathsf{K}^{n \times n}$ is non-singular. Wiedemann (1986) presents a Las Vegas randomized algorithm that will solve $Ax = b$ for x, where $x, b \in \mathsf{K}^{n \times 1}$, that requires an expected number of $O(n)$ applications of the black box for A and $O(n^2)$ additional arithmetic operations in K. For the singular case, Wiedemann shows how to compute a random vector in the solution manifold in the same expected number of steps. Furthermore, Wiedemann gives a Las Vegas randomized algorithm that computes the determinant of A in $O(n)$ expected number applications of the black box for A and an additional $O(n^2 \log(n))$ arithmetic steps in K.

$A \in \mathsf{K}^{n \times n}$
K a field of cardinality $\geq 50n^2 \log(n)$

Figure 3: Black box representation of a matrix.

It is again important to realize that Wiedemann's results generalize sparse matrix representation. Clearly, for sparse matrices, the product Ay may only take $O(n)$ arithmetic steps in K. In that case the methods improve on any $O(n^3)$ algorithms for system solving. However, the algorithms also apply to dense matrices with special structure, such as Toeplitz, Vandermonde, or resultant matrices. In (Canny et al. 1989), e.g., we present an efficient algorithm to multiply a Macaulay (1916) resultant matrix and a vector, and thus obtain by Wiedemann's algorithms the most efficient solution known for computing the Macaulay resultant itself. The black box representation of a matrix that allows its multiplication with a given vector is of course a ubiquitous concept in numerical sparse linear system solving, such as in the Lanczos and conjugate gradient methods for solving positive definite sparse linear systems (see (Golub and van Loan 1987)). Wiedemann's algorithms are distinguished from these methods in that they are algebraic and compute an exact answer in contrast to a numerical approximation of the solution.

Notation: We use the symbols := and =: to define new mathematical objects (the new quantities being on the side of the colon), and we use the symbols $\leftarrow$ and $\rightarrow$ as assignment operators in program code. Furthermore, $\deg(f)$ always denotes the total degree of the multivariate polynomial f.

2. Factorization of Polynomials in Black Box Representation

We now present the algorithm that takes a black box for a multivariate polynomial and produces a program that will evaluate all individual irreducible factors at arbitrary values for the variables.

Algorithm *Black Box Polynomial Factorization*

Input: A black box that evaluates $f(x_1, \ldots, x_n) \in \mathsf{K}[x_1, \ldots, x_n]$, where K is a field of characteristic zero. We also assume that we have an effective polynomial factorization algorithm for $\mathsf{K}[y]$ (see Figure 1). Furthermore, a failure probability $\epsilon \ll 1$ is input.

Output: Assume that

$$f(x_1, \ldots, x_n) = \prod_{i=1}^{r} h_i(x_1, \ldots, x_n)^{e_i}, \quad e_i \geq 1,$$

is the factorization of f into irreducibles. First we output integers $\bar{e}_1, \ldots, \bar{e}_n$ such that with probability no less than $1 - \epsilon$, $e_i = \bar{e}_i$ for all $1 \leq i \leq r$.

We also output a program (see Figure 2) that makes oracle calls to the black box for f and has with probability at least $1 - \epsilon$ the following property. The program accepts as input n arbitrary elements $p_1, \ldots, p_n$ in K. It returns the values

$$h_1(p_1, \ldots, p_n), h_2(p_1, \ldots, p_n), \ldots, h_r(p_1, \ldots, p_n) \in \mathsf{K}.$$

Notice that the h_i are determined only up to a multiple in K. The constructed program once and for all chooses an associate for each factor h_i and, for repeated invocations with different arguments, returns the value of that associate. Notice also that the failure probability applies to the construction and not to the execution of the program. That is, with probability at least $1 - \epsilon$ the output program is correct; a correct program *always* produces the true values of the factors.

Step 1: *Pick random field elements*

$$a_2, \ldots, a_n, b_1, \ldots, b_n, c_1, c_3, \ldots, c_n$$

from a sufficiently large finite subset $R \subset \mathsf{K}$. We will give the cardinality of this set in relation to $\deg(f)$ and ϵ in the statement of theorem 1 below.

By standard interpolation compute

$$f_2(X_1, X_2) := f(X_1 + c_1X_2 + b_1, a_2X_1 + X_2 + b_2,\\ a_3X_1 + c_3X_2 + b_3, \ldots, a_nX_1 + c_nX_2 + b_n).$$

The interpolation algorithm needs to know $\deg(f)$. Either $\deg(f)$ or an upper bound is also supplied as input, or it can be probabilistically guessed beforehand as follows (see also (Kaltofen 1988)). First, we will interpolate $f_1(X_1) := f_2(X_1, 0)$. With high probability, depending on the cardinality of R, we have $\deg(f) = \deg(f_1)$. To find $\deg(f_1)$ we compute a succession of polynomials $f_1^{(d)}$ for $d = 1, 2, 3, \ldots$ until $f_1^{(d)} = f_1$, where $f_1^{(d)}(X_1)$ is the interpolate at $X_1 = 0, 1, \ldots, d$ of $f_1(X_1)$. We test whether $f_1^{(d)} = f_1$ by evaluation at a random $A \in R$: if $f_1^{(d)} = f_1(A)$ then we declare $f_1 = f_1^{(d)}$, and $\deg(f_1) = d$. As discussed in (Freeman et al 1986, procedure **StraightDegree**), the $f_1^{(d)}$ can be computed incrementally by using a divided difference table.*

The probability that this method determines $\deg(f)$ correctly depends on the relation of $\deg(f)$ to the cardinality of the set R, from which the random elements are taken. Since the algorithm chooses the set R, say integers of no more than l bits, where l is appropriately selected, some upper bound for $\deg(f)$ must be known to the algorithm beforehand. This is unavoidable since the black-box could compute a polynomial in which all l-bit integer translation points are 'unlucky'. For example, for all such integers, the projected polynomial f_2 might be the zero polynomial, hence unusable. In any case, the failure probability is proportional to $\deg(f)/2^l$, which means that the degree of f would have to be very large in order for the guessing procedure to be unreliable. Hence this problem of having to be given information about the magnitude of $\deg(f)$ is more a theoretical than a practical issue.

Step 2: *Factor $f_2(X_1, X_2)$ over $\mathsf{K}[X_1, X_2]$ into irreducibles,*

$$f_2(X_1, X_2) =: \prod_{i=1}^{\tilde{r}} g_{2,i}(X_1, X_2)^{\tilde{e}_i}, \quad \tilde{e}_i \geq 1.$$

*An even simpler method of guessing $\deg(f_1)$ has been proposed by R. Liebling. Consider the following difference operator on $\mathsf{K}[X_1]$:

$$\Delta_0[\![f_1]\!] := f_1, \quad \Delta_{i+1}[\![f_1]\!](X_1) := \Delta_i[\![f_1]\!](X_1 + 1) - \Delta_i[\![f_1]\!](X_1) \text{ for all } i \geq 0.$$

For a random $A \in R$ we compute

$$\Delta_0[\![f_1]\!](A), \Delta_1[\![f_1]\!](A), \Delta_2[\![f_1]\!](A), \ldots$$

until we get $\Delta_{d+1}[\![f_1]\!](A) = 0$. Notice that a new $\Delta_{i+1}[\![f_1]\!](A)$ can be found by keeping the values of $\Delta_j[\![f_1]\!](A + i - j)$ with $0 \leq j \leq i$, and updating this fringe of a triangular difference table by the iteration: *For* $j \leftarrow 1, \ldots, i+1$ *Do:*

$$\Delta_j[\![f_1]\!](A + i + 1 - j) \leftarrow \Delta_{j-1}[\![f_1]\!](A + i + 2 - j) - \Delta_{j-1}[\![f_1]\!](A + i + 1 - j).$$

Notice that this bivariate factorization problem is reducible to a univariate one, in polynomially many field operations (Kaltofen 1985a). Moreover, by virtue of the effective Hilbert irreducibility theorem (Kaltofen 1985b and 1989), for a set R of cardinality $2^{\deg(f)^{1+o(1)}}/\epsilon$ the factor degree patterns of f and f_2 agree with probability no less than $1-\epsilon$, that is, $d = \deg(f)$, $r = \tilde{r}$, and for all $1 \le i \le r$, $e_i = \tilde{e}_i$. Let us suppose that all this is true. Otherwise an incorrect output program will be produced.

Step 3: *Assign*

$$g_{1,i}(X_1) := g_{2,i}(X_1, 0) \quad \textit{for all } 1 \le i \le r.$$

Check whether $\mathrm{GCD}(g_{1,i}, g_{1,j}) = 1$ *for all* $1 \le i < j \le r$, *and whether* $\deg(g_{2,i}) = \deg(g_{1,i})$ *for all* $1 \le i \le r$. *If not, return "failure".* In that case we know that the random points were unlucky by zeroing one of several possible polynomials whose zeroes are to be avoided.

We now have, with high probability,

$$f_1(X_1) = f(X_1 + b_1, a_2X_1 + b_2, \ldots, a_nX_1 + b_n) = \prod_{i=1}^{r} g_{1,i}(X_1)^{e_i}, \tag{1}$$

where each $g_{1,i}$ is an image of an h_i. We first map from multi- to bivariate polynomials in step 2 and then from bi- to univariate polynomials in step 3 because of a lack of a univariate Hilbert irreducibility theorem for K. By that we mean a theorem that establishes the existence of a set of substitutions for the variables that map all irreducible factors of f to univariate irreducible polynomials. For general fields K such a theorem does not exist; e.g., if K is algebraically closed, f_1 will always factor into linear factors over K, even if f is irreducible. However, many other fields, such as finite algebraic extensions of Q, yield a univariate Hilbert irreducibility theorem. Nonetheless, we know of no computationally effective versions, so for theoretical reasons we still must apply the bivariate counterpart. However, in practice it is unlikely that f violates the conclusion of such a theorem, which is that the factor degree pattern of f_1 with high probability agrees with that of f. Therefore, in practice, for such fields it is not unreasonable to map f directly to f_1 and then factor f_1 over $\mathsf{K}[X_1]$.

We will use $g_{1,i}$ to uniquely enumerate the factors of f. Our associate choices (see the output specifications above) will satisfy

$$h_i(X_1 + b_1, a_2X_1 + b_2, a_3X_1 + b_3, \ldots, a_nX_1 + b_n) = g_{1,i}(X_1).$$

Step 4: This step constructs the program for evaluation of the h_i at $p_1, \ldots, p_n$ as described in the output specifications. First, the information computed so far is 'hardwired' into that program. Then the following steps A, B, and C, are appended to the program.

Step A: *By standard interpolation using the determined* $\deg(f)$ *compute*

$$\begin{aligned}\bar{f}(X_1, Y) := f(X_1 + b_1, Y(p_2 - a_2(p_1 - b_1) - b_2) + a_2X_1 + b_2,\\ \ldots, Y(p_n - a_n(p_1 - b_1) - b_n) + a_nX_1 + b_n).\end{aligned}$$

Notice that

$$\bar{f}(p_1 - b_1, 1) = f(p_1, \ldots, p_n) \quad \text{and} \quad \bar{f}(X_1, 0) = f_1(X_1).$$

Step B: This step computes the factorization

$$\bar{f}(X_1, Y) = \prod_{i=1}^{r} \bar{g}_i(X, Y)^{e_i} \quad \text{with } \bar{g}_i(X_1, 0) = g_{1,i}(X_1) \text{ for all } 1 \le i \le r.$$

Note that since the polynomials $g_{1,i}$ are not necessarily irreducible, neither might be the $\bar{g}_i$. One way to obtain the factorization would be to factor $\bar{f}(X_1, Y)$ over $\mathsf{K}[X_1, Y]$ into irreducible factors, evaluate these factors at $Y = 0$, and then check which of the $g_{1,i}$ the corresponding univariate images divide. Finally, we can compute the product of those irreducible factors of $\bar{f}$ whose univariate images divide a given $g_{1,i}$ and adjust by a scalar multiplier to obtain $\bar{g}_i$. However, this would require the bivariate factorization of $\bar{f}$, and there is a much more efficient solution, namely by Hensel lifting the factorization (1) to one for $\bar{f}$. We will not describe the lifting algorithm in detail, but instead refer to (Musser 1976) and (von zur Gathen and Kaltofen 1985). The latter article explains that one can in our setting lift with the multiplicities e_i.

By Hensel lifting (1) obtain a factorization

$$\prod_{i=1}^{r} \bar{g}_i(X_1, Y)^{e_i} \equiv \bar{f}(X_1, Y) \pmod{Y^{d+1}}, \quad \deg_y(\bar{g}_i) \le d.$$

For all $1 \le i \le r$ *test, whether* $\bar{g}_i$ *divides* $\bar{f}$. *If at least one test fails, return "program incorrect".* In that case we have discovered that the factor degree patterns of f and f_2 must disagree.

Step C: *For* $i \leftarrow 1, \ldots, r$ *Do:*

Return $\bar{g}_i(p_1 - b_1, 1)$ *as the value* $h_i(p_1, \ldots, p_n)$, *the value of the* i*-th irreducible factor.*

□

The following two theorems discuss the complexity of this algorithm. As mentioned at the end of step 1, any failure estimates are dependent on the degree of f, which is not supplied as an input parameter and cannot be synthetically produced from the black box alone. Therefore, the failure probability can only be guaranteed if the set R in step 1 was selected of at least the stated cardinality.

THEOREM 1. *The Black Box Polynomial Factorization algorithm can construct its output program in polynomially many arithmetic steps as a function of* n *and* $\deg(f)$ *and an additional single polynomial factorization in* $\mathsf{K}[X_1]$. *It requires* $O(\deg(f)^2)$ *oracle calls to the black box for* f. *If the cardinality of the set* R *in step 1 is chosen*

$$\operatorname{card}(R) \ge 6 \deg(f)\, 2^{\deg(f)} / \epsilon,$$

then the algorithm succeeds with probability no less than $1 - \epsilon$ *and the resulting program will always correctly evaluate all irreducible factors of* f. *That program in turn can be executed in polynomially many arithmetic steps and* $O(\deg(f)^2)$ *calls to the black box for* f. *For* $\mathsf{K} = \mathsf{Q}(\vartheta)$, *where* ϑ *is an algebraic number given by its minimal polynomial* $\varphi(z) \in \mathsf{Q}[z]$, *that is,* K *is isomorphic to* $\mathsf{Q}[z]/(\varphi(z))$, *the timings in terms of bit complexity are polynomial in* n, $\deg(f)$, *and the additional parameters:* $\log(1/\epsilon)$, *and the coefficient sizes of* f *and* φ; *this under the provision that the set* R *is chosen as the first* $\operatorname{card}(R)$ *positive integers. Furthermore, all evaluations of the black box for* f *within the algorithm are on input values of polynomial size as a function of the stated parameters.*

Proof. The statements on the run time of both the algorithm and the returned program are easily verified. First, we need to interpolate the bivariate polynomials f_2 and $\bar{f}$, both of degree at most $\deg(f)$ (if unlucky elements are chosen in step 1, the degrees might be less). Clearly, each bivariate interpolation requires $O(\deg(f)^2)$ values of f. The algorithm then needs to factor f_2, which can be accomplished for $\mathsf{K} = \mathsf{Q}[z]/(\varphi(z))$ in binary polynomial time in $\deg(f_2)$ and the coefficient size of f_2 (Landau 1985). Notice that one needs to make an appropriate definition for the coefficient size of f in order to guarantee that the coefficient size of f_2 stays polynomially bounded. One such choice is the combined coefficient size of f as defined in Kaltofen (1989), §4, page 393. The dominating additional work of the output program is step B, which essentially is a uni- to bivariate lifting problem that certainly can be accomplished in $O(\deg(f)^4)$ arithmetic operations.

It remains to analyze the failure probabilities. An incorrect program is output if and only if the factor degree patterns of f and f_2 disagree. This happens when $\deg(f_2) < \deg(f)$, or when some h_i maps to a reducible polynomial in $\mathsf{K}[X_1, X_2]$, or when two of the h_i's map to associated irreducible polynomials in $\mathsf{K}[X_1, X_2]$. By theorem 5.2 in (Kaltofen 1989) (or as explained in the proof of theorem 6.1 (ibid.)), this happens with probability less than

$$(\deg(f) + 4\deg(f)\, 2^{\deg(f)} + \deg(f)^3)/\mathrm{card}(R).$$

Furthermore, if we use the procedure described in step 1 for guessing the degree of f, we may pick a value A that leads to an agreement $f_1(A) = f_1^{(d)}(A)$ for $d < \deg(f)$ with probability no more than $\deg(f)^2/\mathrm{card}(R)$. Thus, by choosing $\mathrm{card}(R) > 6\deg(f)\, 2^{\deg(f)}/\epsilon$ we can guarantee that the output program is correct (including that step B will never report an error) with probability no less than $1 - \epsilon$.

Finally, we need to estimate the probability that step 3 reports failure. Such circumstances correspond to failure in step F of the Factorization algorithm in (Kaltofen 1989), and is treated in the fourth case of failure in the proof of theorem 6.1 (ibid.). It is shown that the probability of these events can be bounded from above by $\deg(f)^2/\mathrm{card}(R)$, which for the given cardinality of R certainly lies below ϵ. □

Theorem 1 has several applications to computing the sparse factorization of a multivariate polynomial given by a black box. For instance, we get the following theorem of theoretical interest, which establishes that several sparse factorization questions as belonging to the complexity class randomized $\mathcal{NC}$ for parallel computation (see (Cook 1985)); it also solves an open problem in (von zur Gathen and Kaltofen 1985).

THEOREM 2. *For $\mathsf{K} = \mathsf{Q}(\vartheta)$ as in theorem 1, we have a Monte Carlo $\mathcal{NC}$-reduction from the problem of computing all sparse factors with no more than t terms to the problem of factoring univariate polynomials over Q. This reduction is of polynomial binary operation count and poly-logarithmic depth in terms of n, $\deg(f)$, t, and the coefficient size of f and φ. It requires that the black box calls account for polynomially many Boolean operations and poly-logarithmic depth. In particular, under this assumption for $\mathsf{K} = \mathsf{Q}$ the problem of computing all sparse factors of a fixed degree with no more than t terms is in the complexity class Monte Carlo $\mathcal{NC}$.*

Proof. Constructing and evaluating the output program of the Black Box Polynomial Factorization algorithm is basically interpolating bivariate images of the input black box and bivariate factorization. Bivariate factorization over $\mathsf{K} = \mathsf{Q}[z]/(\varphi(z))$ is $\mathcal{NC}$ reducible to univariate factorization over Q by the reduction from bivariate to univariate factorization over K in (Kaltofen 1985a) and the reduction from factoring univariate polynomials over algebraic number fields to factoring univariate polynomials over the rationals in (Trager 1976). From the program for all factors we can deduce the sparse factors in Monte Carlo $\mathcal{NC}$ by the algorithm of Ben-Or and Tiwari (1988). Note that although the Ben-Or and Tiwari algorithm is deterministic, we must check which of the sparse interpolating polynomials with t or fewer non-zero monomials are correct because the Ben-Or and Tiwari algorithm can yield an incorrect sparse interpolating polynomial for a dense factor. The verification can be done in a Monte Carlo fashion by evaluating both the program for the factor and the computed sparse candidate polynomial at a random point and by comparing the resulting algebraic numbers.

We finally discuss the statement on factors of fixed degree. In step 2 one can in parallel compute those bivariate integral factors that have fixed degree by lifting univariate factors modulo a prime number first to integral factors and then to bivariate factors as in the standard Berlekamp-Hensel method (Kaltofen 1982). The modular univariate factors can be found in parallel by a variation of the Berlekamp algorithm (Gathen 1984). Finally, both the true bivariate factors as well as the true univariate integral factors are recoverable within the complexity $\mathcal{NC}$ because there are only a polynomial number of lifted factor combinations to be tested, since the degree of the product candidate polynomial is fixed. □

We shall conclude this section with a different interpretation of our idea, which was actually conceived after the method had been discovered. One can view steps B and step C in the algorithm to fit the homotopy continuation paradigm for numerically solving a system of equations (see, for instance, (Drexler 1977), (Garcia and Zangwill 1979), (Li, Sauer, and Yorke 1988), and (Zulehner 1988)). Assume that one wants to solve the system of equations $\vec{F}(\vec{x}) = 0$. One parameterizes the indeterminates with a single new variable Y and considers the system

$$\vec{H}(\vec{X}(Y)) = (1 - Y)\vec{G}(\vec{X}(Y)) + Y\vec{F}(\vec{X}(Y)),$$

where $\vec{X}(0)$ is a solution to the specially selected system $\vec{G}(\vec{x}) = 0$ and $\vec{X}(1)$ is sought as a solution of $\vec{H}(\vec{X}(1)) = 0$. In order to obtain $\vec{X}(1)$ one continously deforms $\vec{X}(0)$ by letting the parameter Y go from 0 to 1. Note that in the numerical homotopy methods, the standard parameterization described for the variables is simply $\vec{X}(Y) = \vec{x}$, that is the systems $\vec{G}$ and $\vec{F}$ stay invariant during the deformation.

In our case the problem $\vec{G}(\vec{x}) = 0$ is the problem of factoring f_1, and the problem $\vec{F}(\vec{x}) = 0$ is the problem of finding the corresponding factorization of $\bar{f}(X_1, 1)$. The former problem does not depend on the input $p_1, \ldots, p_n$, whereas the latter does. The Hensel lifting method referred to in step B exactly corresponds to deforming the solution of $\vec{G}(\vec{x}) = 0$ to one for $\vec{H}(\vec{X}(Y))$, although it should be noted that our deformation deforms with a discrete valuation. Since $\deg_Y(\bar{f})$ is finite, this deformation terminates with the exact solution, and evaluation at $Y = 1$ is possible.

3. Black Box GCDs and Separation of Numerators from Denominators

We now discuss two more results concerning the construction of evaluation programs from black boxes. The first is the construction of a program for the evaluation of the greatest common divisor of multivariate polynomials, the second takes a black box for a multivariate rational function and will individually evaluate its numerator and denominator. As in the factorization algorithm, the resulting programs are, with high probability, universally correct: they will return the correct values for all inputs. This requirement makes the construction more difficult. In both the GCD and the separation result we will employ the same idea that we already used in the factorization algorithm above.

We first discuss the computation of GCDs and shall give only a brief explanation. Assume that for $i = 1, \ldots, r$, $r \geq 2$, the non-zero polynomials $f_i(x_1, \ldots, x_n) \in \mathsf{K}[x_1, \ldots, x_n]$ are given by a black box. For field elements $p_1, \ldots, p_n$ we wish to have $g(p_1, \ldots, p_n)$, where $g = \mathrm{GCD}(f_1, \ldots, f_r)$ and we again fix the associate. First we choose random field elements

$$a_2, \ldots, a_n, b_2, \ldots, b_n, c_2, \ldots, c_n.$$

Now with high probability for the univariate image of the GCD,

$$g_1(X) := g(X, a_2X + b_2, \ldots, a_nX + b_n) \in \mathsf{K}[X],$$

we have $\deg(g_1) = \deg(g)$, that is we can normalize g such that g_1 is monic. Furthermore, with high probability (see (Kaltofen 1988, Theorem 6.2)), we have

$$g_1 = \mathrm{GCD}(f_1(X, a_2X + b_2, \ldots, a_nX + b_n), \sum_{i=2}^{r} c_i f_i(X, a_2X + b_2, \ldots, a_nX + b_n)).$$

By polynomial interpolation we now compute the two bivariate polynomials

$$\bar{f}_0(X,Y) := \sum_{i=2}^{r} c_i \bar{f}_i(X,Y) \quad \text{and} \quad \bar{f}_1(X,Y),$$

where the $\bar{}$ operator is for any $h \in \mathsf{K}[x_1, \ldots, x_n]$ the projection

$$\bar{h}(X,Y) := h(X, Y(p_2 - a_2p_1 - b_2) + a_2X + b_2, \ldots, Y(p_n - a_np_1 - b_n) + a_nX + b_n).$$

Next, we compute $\bar{g}(X,Y)$ as the $\mathrm{GCD}(\bar{f}_0(X,Y), \bar{f}_1(X,Y))$, where $\bar{g}$ is normalized by making the leading coefficient $\mathrm{ldcf}_X(\bar{g}) = 1$. We actually get $\bar{g}$ and the normalization amounts to a scalar division if g_1 does not drop in degree compared with the degree of g, that with high probability. Finally, we return as $g(p_1, \ldots, p_n)$ the value $\bar{g}(p_1, 1)$.

Even though this is more or less the same idea as the one in the factorization algorithm, the transcendental Y is used for a somewhat different reason here. The problem is that one cannot guarantee that the GCD of the polynomials $\bar{f}_0(X,1)$ and $\bar{f}_1(X,1)$ is an image of g, and our algorithm is to produce a black box which is correct with high probability; correct means that it evaluates the GCD at all points. It is therefore intriguing to compare the result with the polynomial GCD algorithm for the straight-line representation (Kaltofen 1988, §6). In fact, a similar problem arises for the solution presented there, but in a somewhat more implicit way. Although the resulting straight-line program for the GCD is derived by a univariate polynomial remainder sequence (see step R in the cited algorithm), that program will contain divisions. Therefore, it cannot be evaluated at all points in the straight-forward manner. One has to first remove the divisions by Strassen's algorithm (see Theorem 7.1 in (Kaltofen 1988)). That again leads a consequent increase in the complexity for computing the values everywhere.

As the last Black Box result in this article we come to the problem of computing the values of the numerator and the denominator of a rational function. A certain subproblem, that of a mixed radix Padé approximation, will be used in both the algorithm constructing the resulting program and in the program itself. For clarity, we shall present the algorithm for this subproblem below the main algorithm.

Algorithm *Black Box Numerator and Denominator*

Input: A black box that evaluates $f(x_1, \ldots, x_n)/g(x_1, \ldots, x_n) \in \mathsf{K}(x_1, \ldots, x_n)$, where K is a field of characteristic 0, $f, g \in \mathsf{K}[x_1, \ldots, x_n]$ with $\mathrm{GCD}(f, g) = 1$. The black box returns ∞ if one evaluates at a root of g. Furthermore a failure probability $\epsilon \ll 1$ is input. Without the knowledge of a bound $\bar{e} \geq \deg(g)$, this algorithm might with positive probability loop forever. If this is not desired, a degree bound $\bar{e}$ is also required as input.

Output: A program (see Figure 4) that makes oracle calls to the black box for f/g and has with probability at least $1 - \epsilon$ the following property. The program accepts n field elements $p_1, \ldots, p_n$. It returns the values of

$$f(p_1, \ldots, p_n) \quad \text{and} \quad g(p_1, \ldots, p_n).$$

As in the Black Box Polynomial Factorization algorithm, certain associates for f and g are chosen once and for all and the values of these associates will be returned.

Notice that the program will be able to return the value of $f(p_1, \ldots, p_n)$ even if $g(p_1, \ldots, p_n) = 0$. This apparent paradox—the black box for f/g does not supply any information on f when probed at $p_1, \ldots, p_n$—is resolved by deducing the value of $f(p_1, \ldots, p_n)$ from values at other points that are not zeros of g.

Step 1: This step introduces a random projection of f and g into $\mathsf{K}[X]$ that with high probability preserves their relative primality and the degree of g.
From a sufficiently large finite set $R \subset \mathsf{K}$ *randomly select* a_i, b_i, $2 \leq i \leq n$. In place of x_i we will evaluate at $a_i X + b_i$ for $2 \leq i \leq n$, such that for

$$f_1(X) := f(X, a_2 X + b_2, \ldots, a_n X + b_n)$$

and

$$g_1(X) := g(X, a_2 X + b_2, \ldots, a_n X + b_n)$$

we have $\mathrm{GCD}(f_1, g_1) = 1$ over $\mathsf{K}[X]$ and $\deg(g_1) = \deg(g)$: depending on the cardinality of R this is true with high probability. Notice that the degree condition on g_1 implies that $g_1 \neq 0$.

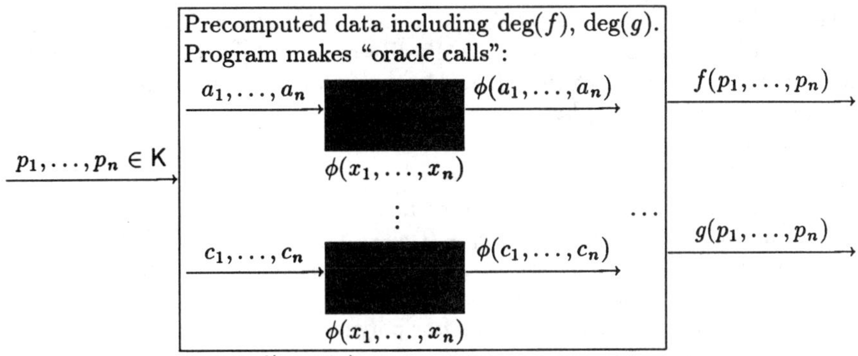

$$\phi(x_1, \ldots, x_n) = \frac{f(x_1, \ldots, x_n)}{g(x_1, \ldots, x_n)}, f, g \in \mathsf{K}[x_1, \ldots, x_n], \mathrm{GCD}(f, g) = 1.$$

Figure 4: The program for evaluating the numerator and denominator of a black box rational function.

Step 2: This step probabilistically computes the degrees of f and g. In order to prevent an infinite loop in the case that $g_1 = 0$ we assume that one is given an additional bound $\bar{e} \geq \deg(g)$.

For $\delta \leftarrow 0, 1, 2, \ldots$ *Do:*

Call the algorithm Special Padé Approximation below with the degree bounds $d = e = \delta$ *and the linear substitutes* $a_2 X + b_2, \ldots, a_n X + b_n$. *Denote the two obtained polynomials by* $f_1^{(\delta)}, g_1^{(\delta)} \in \mathsf{K}[X]$. If both polynomials are 0, then either the projection sent g into the zero polynomial or $\delta < \deg(g)$ together with additional conditions on g_1 (see the output specifications of the Special Padé Approximation Algorithm below).

If $f_1^{(\delta)} = g_1^{(\delta)} = 0$ *increment* δ *by one and continue the For loop, unless* $\delta = \bar{e}$, *in which case return to step 1.* In this exceptional case $g_1 = 0$, provided that $\bar{e}$ is a true bound for $\deg(g)$. Therefore a new random projection must be tried.

If $g_1^{(\delta)} \neq 0$, *pick another random element* $A \in R$ *and compare* $(f/g)(A, a_2 A + b_2, \ldots, a_n A + b_n)$ *computed by the input black box with the value of* $f_1^{(\delta)}(A)/g_1^{(\delta)}(A)$. *If both values are the same and are not* ∞ *then declare* $\deg(f)$ *to be* $\deg(f_1^{(\delta)}) \to d$ *and* $\deg(g)$ *to be* $\deg(g_1^{(\delta)}) \to e$, *and go to step 3. If both values are* ∞, *pick a new distinct random* A *and repeat the comparison. Otherwise continue the For loop with the next* δ.

Step 3: This step constructs the program described in the output specifications. It will permanently use the random elements chosen in step 1 and the degrees computed in step 2. The idea is to compute

$$\bar{f}(X, Y) := f(X, a_2 X + b_2 + Y(p_2 - a_2 p_1 - b_1), \ldots, a_n X + b_n + Y(p_n - a_n p_1 - b_n))$$

and

$$\bar{g}(X, Y) := g(X, a_2 X + b_2 + Y(p_2 - a_2 p_1 - b_1), \ldots, a_n X + b_n + Y(p_n - a_n p_1 - b_n)).$$

For $Y = 0$ we already have computed these polynomials in step 2. Here we compute $\bar{f}_j(X) := \bar{f}(X, i_j)$ and $\bar{g}_j(X) := \bar{g}(X, i_j)$ for sufficiently many distinct integral values i_j, that we can interpolate with respect to Y. However, in order to use the special Padé approximation for the computation of $\bar{f}_j$ and $\bar{g}_j$ these two images must be relatively prime. Therefore, certain values for i_j have to be discarded.

Step A: $j \leftarrow 0$.

For $I \leftarrow 1, 2, \ldots$ *Do:*

Call the algorithm Special Padé Approximation below with the degree bounds d, e *and the linear substitutions* $a_i X + b_i + I(p_i - a_i p_1 - b_i)$ *for* $2 \le i \le n$. *If the degrees of the returned polynomials are below d and e respectively, continue the For loop with the next* I. *Otherwise, set* $i_{j+1} \leftarrow I$, $\bar{f}_{j+1}(X)$ *to the numerator polynomial returned, and* $\bar{g}_{j+1}(X)$ *to the denominator polynomial returned. Increment* j *by* 1. *If* $j = \max(d, e) + 1$ *then exit the loop. Otherwise continue the For loop with the next* I.

As we will show below, if the substitutions of step 1 are lucky the above loop will discard at most $2de$ values of I. Thus, if more values are unusable the entire program is incorrect. Of course, this should be tested for and diagnosed, instead of getting into a possible infinite loop in such an unlucky case.

Step B: *From the polynomials* $\bar{f}_j, \bar{g}_j$ *compute, by interpolation, candidates for the polynomials* $\bar{f}(X, Y), \bar{g}(X, Y)$.

Step C: *Return* $\bar{f}(p_1, 1)$ *and* $\bar{g}(p_1, 1)$ *as the values of* $f(p_1, \ldots, p_n)$ *and* $g(p_1, \ldots, p_n)$ *respectively.* □

Algorithm *Special Padé Approximation*

Input: The same black box for f/g that is input of the Numerator and Denominator algorithm. Furthermore, two degree bounds d, e and linear forms $u_2 X + v_2, \ldots, u_n X + v_n$ with $u_i, v_i \in \mathsf{K}$ for $2 \le i \le n$.

Output: Two polynomials $\hat{f}(X), \hat{g}(X) \in \mathsf{K}[X]$ with the following property. The algorithm has chosen $d + e + 1$ integers

$$1 \le i_1 < i_2 < \cdots < i_{d+e+1}$$

such that for all $1 \le k \le d + e + 1$,

$$\frac{\hat{f}(i_k)}{\hat{g}(i_k)} = \frac{f}{g}(i_k, u_2 i_k + v_2, \ldots, u_n i_k + v_n), \tag{2}$$

and such that

$$\deg(\hat{f}) \le d, \quad \deg(\hat{g}) \le e, \quad \mathrm{ldcf}(\hat{g}) = 1.$$

If the right-hand side of (2) is not ∞ for all k, then such $\hat{f}, \hat{g}$ can be determined (see lemma 1 below). Furthermore, if $d \ge \deg(f)$ and $e \ge \deg(g)$, the algorithm will return relatively prime polynomials $\hat{f}, \hat{g}$. (Refer also to the remark following the algorithm description.) No appropriate i_k can be selected if either

$$0 = g(X, u_2 X + v_2, \ldots, u_n X + v_n) =: \tilde{g}(X),$$

or if $\deg(g) > e$ and more than e of the integers j in $1 \le j \le d + 2e + 1$ are roots of $\tilde{g}(X)$. In these exceptional cases two zero polynomials are returned.

Step I: We first find, if possible, a set of $d+e+1$ integers $1 \le i_1 < i_2 < \cdots < i_{d+e+1} \le d+2e+1$ that are not roots of $\tilde{g}(X) = g(X, u_2X + v_2, \ldots, u_nX + v_n) \in \mathsf{K}[X]$.
$k \leftarrow 0$.
For $J \leftarrow 1, \ldots, d+2e+1$ *Do:*
By evaluation of the black box for f/g *compute*

$$A_{k+1} \leftarrow \frac{f}{g}(J, u_2J + v_2, \ldots, u_nJ + v_n).$$

If $A_{k+1} \ne \infty$ *then* $i_{k+1} \leftarrow J$. *Increment* k *by* 1 *and if* $k = d+e+1$ *exit loop.*
At this point we have tried at most $d+2e+1$ values for J.
If $k < d+e+1$ *then return two zero polynomials for* $\hat{f}$ *and* $\hat{g}$. With

$$\tilde{f}(X) := f(X, u_2X + v_2, \ldots, u_nX + v_n)$$

we now have $A_k = \tilde{f}(i_k)/\tilde{g}(i_k)$ for all $1 \le k \le d+e+1$.

Step II: *By polynomial interpolation compute a polynomial* $h(X) \in \mathsf{K}[X]$ *with* $h(i_k) = A_k$ *for all* $1 \le k \le d+e+1$ *and* $\deg(h) \le d+e$.

Step III: Now we compute polynomials $\hat{f}(X), \hat{g}(X) \in \mathsf{K}[X]$ such that

$$\hat{f}(X) \equiv \hat{g}(X)h(X) \pmod{(X-i_1)\cdots(X-i_{d+e+1})},$$

and

$$\mathrm{GCD}(\hat{f}, \hat{g}) = 1, \quad \deg(\hat{f}) \le d, \quad \deg(\hat{g}) \le e, \quad \mathrm{ldcf}(\hat{g}) = 1.$$

This is accomplished by the extended Euclidean algorithm as follows:
Initialize

$$\begin{aligned} h_0(X) &\leftarrow (X-i_1)\cdots(X-i_{d+e+1}); & g_0(X) &\leftarrow 0; \\ h_1(X) &\leftarrow h(X); & g_1(X) &\leftarrow 1. \end{aligned}$$

For $l \leftarrow 1, 2, \ldots$ *Do:*
If $\deg(h_l) \le d$ *then exit the loop. Otherwise, perform a polynomial division with remainder of* h_{l-1} *and* h_l,

$$\begin{aligned} h_{l+1}(X) &\leftarrow h_{l-1}(X) \pmod{h_l(X)}; \\ q_{l+1}(X) &\leftarrow \big(h_{l-1}(X) - h_{l+1}(X)\big)/h_l(X); \\ g_{l+1}(X) &\leftarrow g_{l-1}(X) - q_{l+1}(X)g_l(X). \end{aligned}$$

Return $\hat{f} \leftarrow \mathrm{ldcf}(g_l)\, h_l$; $\hat{g} \leftarrow g_l/\mathrm{ldcf}(g_l)$. □

We first discuss the auxiliary Special Padé Approximation algorithm in more detail. It can be shown that in the case where $d < \deg(f)$ or $e < \deg(g)$ the fraction $\hat{f}/\hat{g}$ of polynomials satisfying the output conditions is uniquely determined (see, for instance, (Knuth 1981, Exercise 4.6.1–26)). However, since we want to avoid the GCD reduction to relative prime outputs at the end of the algorithm, and because we do not need this fact in step 2, we will not provide a proof. Nonetheless, all other needed conditions can be deduced from the following simple lemma, whose content is long known (see, for instance, (Gragg 1972)).

LEMMA 1. *Let d and e be non-negative integers, and let $F(X), G(X), H(X) \in \mathsf{K}[X]$, K an arbitrary field, $\deg(H) < d+e+1$, $\mathrm{GCD}(F,G) = 1$; furthermore, let i_k, $1 \le k \le d+e+1$, be not necessarily distinct elements in K such that*

$$F \equiv GH \pmod{(X - i_1)\cdots(X - i_{d+e+1})}.$$

Define $h_0(X) := (X - i_1)\cdots(X - i_{d+e+1})$, $\delta_0 := d+e+1$, and $h_1(X) := H(X)$, $\delta_1 := \deg(H)$. Now let $h_l(X), q_l(X) \in \mathsf{K}[X]$ be the l-th remainders and quotients respectively, in the Euclidean polynomial remainder sequence

$$h_{l-2}(X) = q_l(X)h_{l-1}(X) + h_l(X), \quad \delta_l := \deg(h_l) < \delta_{l-1} \text{ for } l \ge 2.$$

In the exceptional case $H = 0$ the sequence is defined to be empty.
Finally, let $f_l(X), g_l(X) \in \mathsf{K}[X]$ be the multipliers in the extended Euclidean scheme $f_l h_0 + g_l h_1 = h_l$, namely,

$$f_0 := g_1 := 1, \quad f_1 := g_0 := 0,$$

$$f_l := f_{l-2} - q_l f_{l-1}, \quad g_l := g_{l-2} - q_l g_{l-1} \quad \text{for } l \ge 2.$$

Then there exists an index j, $1 \le j$, such that $\delta_j \le d < \delta_{j-1}$. For that index we have

$$h_j \equiv g_j H \pmod{(X - i_1)\cdots(X - i_{d+e+1})} \quad \text{and} \quad \deg(g_j) \le e. \tag{3}$$

Furthermore, if $d \ge \deg(F)$ and $e \ge \deg(G)$ then $F = \lambda h_j$, $G = \lambda g_j$ for some $\lambda \in \mathsf{K}$.

Proof. It follows by induction on l that

$$\forall l \ge 2\colon f_l h_0 + g_l h_1 = h_l, \quad \deg(f_l) = \delta_1 - \delta_{l-1}, \quad \deg(g_l) = \delta_0 - \delta_{l-1}.$$

Furthermore, one can easily show that $f_l g_{l-1} - f_{l-1} g_l = \pm 1$, which implies that $\mathrm{GCD}(f_l, g_l) = 1$ (see (Knuth 1981, §4.6.1, Exercise 3)). Since in the Euclidean sequence the remainder 0 eventually appears, an index j can always be found. Now, we have

$$h_j \equiv g_j H \pmod{h_0}, \quad \deg(g_j) = \delta_0 - \delta_{j-1} \le e.$$

This shows that the pair of polynomials h_j, g_j is a solution to (3).
We finally prove uniqueness in case $d \ge \deg(F)$ and $e \ge \deg(G)$. Observing that

$$g_j F - h_j G = (F - GH)\,g_j - (h_j - g_j H)\,G \equiv 0 \pmod{h_0},$$

and the left hand side polynomial has degree less than $d+e+1$, we must have $g_j F = h_j G$. Thus there is a $w(x) \in \mathsf{K}[x]$ with $F = h_j/w$, $G = g_j/w$. Plugging into $h_j = f_j h_0 + g_j h_1$ we infer that then

$$\frac{f_j h_0}{w} = F - GH \equiv 0 \pmod{h_0},$$

which implies that w must divide f_j. From $\mathrm{GCD}(f_j, g_j) = 1$ we thus conclude that $w \in \mathsf{K}$. □

The analysis of the above algorithms now leads us to a theorem very similar in spirit to theorem 1. Notice that only the cardinality of R depends on the failure probability bound ϵ, thus making the arithmetic complexity independent of it, while for properly chosen R (see theorem 1) the binary complexity is polynomially dependent on $\log(1/\epsilon)$ as well. Again as for theorem 1, bounds for the degrees of f and g are required in the estimates. Without any knowledge of degree bounds, one can neither guarantee termination nor estimate the failure probability in any way, since the black box for might evaluate a rational function of astronomical degree for which all choices taken from a selected set R are bad.

THEOREM 3. *Given a bound* $\bar{e} \geq \deg(g)$ *as an additional input, the algorithm Black Box Numerator and Denominator can construct its output program in polynomially many arithmetic steps as a function of* n, $\deg(f)$ *and* $\bar{e}$. *It requires* $O((\deg(f)+\bar{e})^2)$ *oracle calls to the black box for* f/g. *If the cardinality of the set* R *in step 1 is chosen at least*

$$\max\Bigl(2\,(2\deg(f)+1)\deg(g), 3m^2-m\Bigr)\Big/\epsilon, \quad m := \max(\deg(f), \deg(g)),$$

then the resulting program correctly evaluates f *and* g *with probability no less than* $1-\epsilon$. *That program in turn can be executed in polynomially many arithmetic steps and it makes* $O(\deg(f)\deg(g)(\deg(f)+\deg(g)))$ *calls to the black box for* f/g.

Proof. The Special Padé Approximation algorithm takes $O(d+e)$ evaluations of the input black box, where d and e are the supplied degrees. Clearly, step 2 will call this algorithm at most $\max(\deg(f), \bar{e})$ times until the original and the computed rational functions agree at A. In order to deduce the number of calls to the Padé algorithm in step A, we need to estimate how many I can lead to lower degrees. The argument is fairly standard. Consider the Sylvester resultant (see (van der Waerden 1953, §7))

$$\rho_1(Y) := \mathrm{Res}_X(\bar{f}(X,Y), \bar{g}(X,Y)).$$

If f_1 and g_1 are relatively prime images of f and g with $\deg(g_1) = \deg(g)$, then

$$\rho_1(0) = \mathrm{Res}_X(f_1(X), g_1(X)) \neq 0.$$

Furthermore, $\deg(\rho_1) \leq 2de$, and any I with $\rho_1(I) \neq 0$ will satisfy $\mathrm{GCD}(\bar{f}(X,I), \bar{g}(X,I)) = 1$, which proves that in the lucky case at most $2de$ of the I are unusable. Hence the number of evaluations of the input black box that are triggered in step A are at most

$$(2de + \max(d,e)+1)(d+2e+1) = O(\deg(f)\deg(g)(\deg(f)+\deg(g))).$$

It remains to estimate the probability that the returned program is incorrect. Again the argument is standard. Consider the generic leading coefficient

$$g(X, \alpha_2 X + \beta_2, \ldots, \alpha_n X + \beta_n)) =: \tau_1(\alpha_2, \ldots, \alpha_n) X^{\deg(g)} + \cdots,$$

and consider the Sylvester resultant

$$\rho_2(\alpha_2, \ldots, \beta_n) := \mathrm{Res}_X(f(X, \alpha_2 X+\beta_2, \ldots, \alpha_n X+\beta_n), g(X, \alpha_2 X+\beta_2, \ldots, \alpha_n X+\beta_n)).$$

Clearly, $\tau_1(a_2, \ldots, a_n)\rho_2(a_2, \ldots, b_n) \neq 0$ implies $\mathrm{GCD}(f_1, g_1) = 1$ and $\deg(g_1) = \deg(g)$. Since $\deg(\tau_1\rho_2) \leq (2\deg(f)+1)\deg(g)$, the probability of missing this condition in step 1 is no more than

$$\frac{(2\deg(f)+1)\deg(g)}{\mathrm{card}(R)} \leq \frac{\epsilon}{2}.$$

Finally, step 2 must return the correct degrees under the assumption that step 1 picked a lucky evaluation. An element A establishes too low a degree if, despite the fact that $\delta < m = \max(\deg(f), \deg(g))$, it is a zero of the polynomial

$$\tau_2(X) := f_1(X) g_1^{(\delta)}(X) - g_1(X) f_1^{(\delta)}(X) \neq 0.$$

However, $\deg(\tau_2) \leq m + \delta$, so A will be a witness for all $\delta < m$ with probability at least

$$1 - \frac{3m^2 - m}{2\,\mathrm{card}(R)} \geq 1 - \frac{\epsilon}{2}. \quad \square$$

As for theorem 1, when K is an algebraic number field and in particular the rational numbers, the algorithm Black Box Numerator and Denominator is of polynomial-time bit complexity under the usual assumptions. The algorithm also can be used to solve the sparse rational function recovery problem (Ben-Or and Tiwari 1988). One version of that problem assumes that one has term count as well as degree bounds for both the numerator and the denominator. Unlike the sparse polynomial interpolation algorithm in (Ben-Or and Tiwari 1988), our solution uses randomization. However, the final answer can be verified, that is the randomization can be made 'Las Vegas'. Let t be a common term bound for both the actual numerator f and denominator g, and let $\tilde{f}$ and $\tilde{g}$ be the computed sparse numerator and denominator polynomials, respectively. Notice that the sparse interpolation methods might return incorrect polynomials, but they will always have no more than t terms. Therefore $h := f\tilde{g} - \tilde{f}g$ is a polynomial with no more than $2t^2$ terms. Then, if

$$\frac{\tilde{f}(p_1^l, \ldots, p_n^l)}{\tilde{g}(p_1^l, \ldots, p_n^l)} = \frac{f}{g}(p_1^l, \ldots, p_n^l)$$

for distinct primes $p_1, \ldots, p_n$ and for all l with $1 \leq l \leq 2t^2$, then we must have $h = 0$ (Zippel 1990), that is $\tilde{f} = f$ and $\tilde{g} = g$. We have the following theorem.

THEOREM 4. *Given a black box for f/g as in the algorithm Black Box Numerator and Denominator, and given an upper bound t for the number of non-zero monomials in both f and g, we can in expected polynomially many arithmetic steps in* $\deg(f)$, $\deg(g)$, n, *and t compute the sparse representations of f and g.*

Proof. One distinction of this to the previous results is that no degree estimates are needed as inputs. The reason is that the verification procedure discussed above is always correct. Therefore, the algorithm can step through degree estimates and select a set R such that the failure probability is exponentially small. This strategy may lead to an endless run, but the expected value of the run time will be polynomial in the said parameters. □

Notice that this theorem can also be formulated in terms of bit complexity for an algebraic coefficient field $\mathsf{K} = \mathsf{Q}(\vartheta)$ similar to treatment of this special case in theorem 1.

4. Conclusion

Oracle calls in algorithms is a ubiquitous paradigm in complexity reductions such as Turing reductions in recursive function theory, Cook reductions in the theory of $\mathcal{NP}$-completeness, or reductions that relate the matrix multiplication complexity asymptotically to other linear algebra problems such as solving linear systems. Here we have introduced this paradigm to several classical problems on multivariate polynomials. We solved the greatest common divisor, factorization, and numerator/denominator problems.

One application of the black box factorization procedure is that to factor u-resultants (Macaulay 1916). In fact, in that case all factors are known to be linear forms and the algorithm specializes to a method very similar to that in (Canny 1988). An efficient black box for evaluating the u-resultant is discussed in (Canny et al. 1989), and this algorithm, based on Wiedemann's (1986) randomized sparse determinant procedure, is particularly hard to formulate in the straight-line model.

From a pragmatic point of view, the black box model appears to be superior to the straight-line program model because of space requirements. Algorithms acting on straight-line programs usually amplify the length of the result by a factor depending on the degree, whereas the programs making oracle black box calls are of small additional size over the input black boxes. Furthermore, one can adapt most of the results to finite coefficient fields of sufficiently high characteristic and thus avoid costly rational coefficient arithmetic. Moreover, the resulting evaluation programs are especially easy to distribute to asynchronous parallel processors for multiple evaluation in parallel, and perhaps subsequent sparse interpolation. We believe that the black box approach not only provides a simple, but also a practically very efficient solution to, say, the problem of sparse multivariate polynomial factorization.

Acknowledgement: We thank Bobby Caviness, Stephen Cook, Lakshman Yagati, and Prasoon Tiwari for their comments on this work. We especially thank John A. Abbott, David Bayer, and one anonymous referee for their detailed comments to this paper.

Literature Cited

Baur, W. and Strassen, V., "The complexity of partial derivatives," *Theoretical Comp. Sci.* **22**, pp. 317–330 (1983).

Ben-Or, M. and Tiwari, P., "A deterministic algorithm for sparse multivariate polynomial interpolation," *Proc. 20th Annual ACM Symp. Theory Comp.*, pp. 301–309 (1988).

Canny, J., *The Complexity of Robot Motion Planning*; ACM Ph.D. Thesis Award 1987; MIT Press, Cambridge, MA, 1988.

Canny, J., Kaltofen, E., and Lakshman, Yagati, "Solving systems of non-linear polynomial equations faster," *Proc. ACM-SIGSAM 1989 Internat. Symp. Symbolic Algebraic Comput.*, pp. 121–128 (1989).

Cook, S. A., "A taxonomy of problems with fast parallel algorithms," *Inf. Control* **64**, pp. 2–22 (1985).

Drexler, F. J., "Eine Methode zur Berechnung sämtlicher Lösungen von Polynomgleichungssystemen," *Numer. Math.* **29**, pp. 45–58 (1977). (In German.)

Freeman, T. S., Imirzian, G., Kaltofen, E., and Lakshman, Yagati, "DAGWOOD: A system for manipulating polynomials given by straight-line programs," *ACM Trans. Math. Software* **14/3**, pp. 218–240 (1988).

Garcia, C. B. and Zangwill, W. I., "Finding all solutions to polynomial systems and other systems of equations," *Math. Program.* **16**, pp. 159–176 (1979).

von zur Gathen, J., "Parallel algorithms for algebraic problems," *SIAM J. Comp.* **13**, pp. 802–824 (1984).

von zur Gathen, J., "Irreducibility of multivariate polynomials," *J. Comp. System Sci.* **31**, pp. 225–264 (1985).

von zur Gathen, J., "Representations and parallel computations for rational functions," *SIAM J. Comp.* **15**, pp. 432–452 (1986).

von zur Gathen, J. and Kaltofen, E., "Factoring sparse multivariate polynomials," *J. Comp. System Sci.* **31**, pp. 265–287 (1985).

Golub, G. H. and van Loan, C. F., *Matrix Computations*; Johns Hopkins University Press, Baltimore, Maryland, 1987.

Gragg, W. B., "The Padé table and its relation to certain algorithms of numerical analysis," *SIAM Review* **14/1**, pp. 1–62 (1972).

Grigoryev, D. Yu., Karpinski, M., and Singer, M. F., "Fast parallel algorithms for sparse multivariate polynomial interpolation over finite fields," *Tech. Rep.* **8523-CS**, Univ. Bonn, May 1988.

Kaltofen, E., "Polynomial factorization," in *Computer Algebra, 2nd ed.*, edited by B. Buchberger et al.; Springer Verlag, Vienna, pp. 95–113, 1982.

Kaltofen, E., "Polynomial-time reductions from multivariate to bi- and univariate integral polynomial factorization," *SIAM J. Comp.* **14**, pp. 469–489 (1985a).

Kaltofen, E., "Effective Hilbert irreducibility," *Information and Control* **66**, pp. 123–137 (1985b).

Kaltofen, E., "Uniform closure properties of p-computable functions," *Proc. 18th ACM Symp. Theory Comp.*, pp. 330–337 (1986b).

Kaltofen, E., "Greatest common divisors of polynomials given by straight-line programs," *J. ACM* **35/1**, pp. 231–264 (1988).

Kaltofen, E., "Factorization of polynomials given by straight-line programs," in *Randomness in Computation*, Advances in Computing Research **5**, edited by S. Micali; JAI Press Inc., Greenwich, CT., pp. 375–412, 1989.

Kaltofen, E. and Lakshman, Yagati, "Improved sparse multivariate polynomial interpolation algorithms," *Proc. ISSAC '88, Springer Lect. Notes Comput. Sci.* **358**, pp. 467–474 (1988).

Knuth, D. E., *The Art of Programming, Vol. 2, Semi-Numerical Algorithms, Ed. 2*; Addison Wesley, Reading, MA, 1981.

Landau, S., "Factoring polynomials over algebraic number fields," *SIAM J. Comp.* **14**, pp. 184–195 (1985).

Li, T.-Y., Sauer, T., and Yorke, J. A., "Numerically determining solutions of systems of polynomial equations," *AMS Bulletin* **18**/2, pp. 173–177 (1988).

Macaulay, F. S., "Algebraic theory of modular systems," *Cambridge Tracts* **19**, Cambridge, 1916.

Musser, D. R., "Multivariate polynomial factorization," *J. ACM* **22**, pp. 291–308 (1975).

Trager, B. M., "Algebraic factoring and rational function integration," *Proc. 1976 ACM Symp. Symbolic Algebraic Comp.*, pp. 219–228 (1976).

Valiant, L., "Reducibility by algebraic projections," *L'Enseignement mathématique* **28**, pp. 253–268 (1982).

van der Waerden, B. L., *Modern Algebra*; F. Ungar Publ. Co., New York, 1953.

Wiedemann, D., "Solving sparse linear equations over finite fields," *IEEE Trans. Inf. Theory* **IT-32**, pp. 54–62 (1986).

Zippel, R. E., "Interpolating polynomials from their values," *J. Symbolic Comput.*, this issue (1990).

Zulehner, W., "A simple homotopy method for determining all isolated solutions to polynomial systems," *Math. Comp.* **50**/181, pp. 167–177 (1988).

Shortest Division Chains in Imaginary Quadratic Number Fields*

HEINRICH ROLLETSCHEK

Johannes Kepler-University, RISC-Linz,
4040 Linz, AUSTRIA

(*Received* 30 *December* 1987)

Let O_d be the set of algebraic integers in an imaginary quadratic number field $Q[\sqrt{d}]$, $d < 0$, where d is the discriminant of O_d. Consider the Euclidean Algorithm (EA), applied to algebraic integers ξ, $\eta \in O_d$. It consists in computing a sequence of remainders $\rho_0 = \xi$, $\rho_1 = \eta$, ρ_2, ..., $\rho_{n+1} = 0$, where $\rho_{i+1} = \rho_{i-1} - \gamma_i\rho_i$ for algebraic integers $\gamma_i \in O_d$, $i = 1, \ldots, n$. We show that except for $d = -11$ the number of divisions to be carried out is always minimized by choosing each γ_i such that $N(\rho_{i-1} - \gamma_i\rho_i)$, the norm of $\rho_{i-1} - \gamma_i\rho_i$, is minimal. This result has been proven previously in special cases. It also applies to those imaginary quadratic number rings which are not Euclidean; in this case the division chains may be infinite. For $d = -7, -8$ the methods applied so far must be modified somewhat, and for $d = -11$ we provide a counterexample and a theorem which partially answers the question, how shortest division chains can be obtained.

1. Introduction

Given two elements ξ, η of a unique factorization domain (UFD) R, we can attempt to compute their greatest common divisor $\gcd(\xi, \eta)$ by applying the Euclidean algorithm (EA), that is, by computing a sequence of remainders

$$\rho_0 = \xi, \quad \rho_1 = \eta,$$
$$\rho_{i+1} = \rho_{i-1} - \gamma_i\rho_i$$

for appropriate elements $\gamma_i \in R$; if $\rho_{n+1} = 0$ for some n, then $\rho_n = \gcd(\xi, \eta)$. If on the other hand $\rho_{n+1} \neq 0$ for all n, then we say that (EA) fails. This may of course depend on the choice of γ_1, γ_2, ..., and in general it is a non-trivial problem to determine, which choice, if any, accomplishes a remainder $\rho_{n+1} = 0$, and if so, which choice minimizes n, the number of divisions to be carried out.

*This material is based on work supported by the Österreichische Forschungsgemeinschaft under Grant No. 09/0004. A preliminary version of this paper will appear in the *Proceedings of the 1988 Symposium of Symbolic and Algebraic Computation.*

The main result of this paper says that if R is O_d, the set of algebraic integers in an imaginary quadratic number field $\mathrm{Q}[\sqrt{d}]$, $d < 0$, $d \neq -11c^2$ for $c \in \mathrm{Z}$, then n is minimized by choosing each γ_i such that $\mathrm{N}(\rho_{i+1}) = \mathrm{N}(\rho_{i-1} - \gamma_i \rho_i)$ is minimal, where for $\xi \in \mathrm{Q}[\sqrt{d}]$ $\mathrm{N}(\xi)$ is the norm of ξ. This means, of course, that O_{-11} is the only imaginary quadratic domain which, surprisingly, forms an exception, since $\mathrm{Q}[\sqrt{-11c^2}] = \mathrm{Q}[\sqrt{-11}]$ for $c \in \mathrm{Z}$. A slightly weaker result will be shown for this exceptional domain. (Elementary definitions, like that of an algebraic integer, will be reviewed below.)

Things can be formalized by using the notion of ***division chain*** (Cooke (1976)). It will be more convenient for the proof to allow $\rho_0, \rho_1, \ldots$ to be arbitrary elements of $\mathrm{Q}[\sqrt{d}]$, not necessarily algebraic integers. For a sequence $\Sigma = \langle \rho_0, \rho_1, \ldots \rangle$, where $\rho_0, \rho_1, \ldots \in \mathrm{Q}[\sqrt{d}]$, define

$$l(\Sigma) = \begin{cases} n+1, & \text{if } \Sigma = \langle \rho_0, \ldots, \rho_{n+1} \rangle, \\ \infty, & \text{if } \Sigma \text{ is infinite.} \end{cases}$$

Then Σ is a division chain, iff

(i) for each $i = 1, 2, \ldots$ there exists an element $\gamma_i \in O_d$ such that $\rho_{i+1} = \rho_{i-1} - \gamma_i \rho_i$, and
(ii) $\rho_i = 0$ if and only if $l(\Sigma) < \infty$ and $i = l(\Sigma)$.

Thus a division chain Σ is a possible sequence of remainders computed when (EA) is applied to elements $\rho_0, \rho_1 \in \mathrm{Q}[\sqrt{d}]$, and $l(\Sigma) = \infty$ iff (EA) fails.

Consider a division with remainder of ξ by η: $\xi = \gamma\eta + \rho$, where $\gamma \in O_d$, $\xi, \eta, \rho \in \mathrm{Q}[\sqrt{d}]$. If γ is chosen such that $\mathrm{N}(\rho)$ is minimal, then we call the division a minimal remainder-division, ρ a minimal remainder, γ a minimal remainder-quotient of ξ, η. Note that even the minimal remainder of ξ, η is not always uniquely determined. If in a division chain $\Sigma = \langle \rho_0, \rho_1, \ldots \rangle$ ρ_{i+1} is a minimal remainder of ρ_{i-1} and ρ_i for every i, $1 \leq i < l(\Sigma)$, then we call Σ a minimal remainder-division chain. Our result can now be formulated as follows:

THEOREM 1. *If $\Sigma = \langle \rho_0, \rho_1, \rho_2, \ldots \rangle$ is an arbitrary division chain, $\Sigma' = \langle \rho_0, \rho_1, \rho_2', \rho_3', \ldots \rangle$ a minimal remainder-division chain in $\mathrm{Q}[\sqrt{d}]$, $d < 0$, $d \neq -11c^2$ for $c \in \mathrm{Z}$, where Σ and Σ' begin with the same two elements ρ_0, ρ_1, then $l(\Sigma') \leq l(\Sigma)$.*

This theorem applies both for $l(\Sigma) < \infty$ and for $l(\Sigma) = \infty$. Thus if (EA) fails for input (ξ, η) if minimal remainders are used, then it also fails for every other choice of the remainders. Also, the theorem remains valid when O_d is *not* a unique factorization domain (UFD). True, this case is of little practical interest, since (EA) would not be applied in such domains anyway. We include it in our proof, however, since it requires no extra effort. In the case $d \in \{-19, -43, -67, -163\}$, where O_d is a UFD, but not Euclidean with respect to the norm (see below), theorem 1 is of interest in so far, as it provides a new proof for the fact that O_d is not Euclidean with respect to any degree function, which was first shown in Samuel (1971). In fact, a somewhat stronger assertion follows from theorem 1, see Kaltofen/Rolletschek (1985), where the proof for this case was given.

We now discuss, which other cases have been settled so far, and what remains to be done. The proof in Kaltofen/Rolletschek (1985) actually carries over to all discriminants $d \leq -15$ without any modification; this is in fact the easiest case. Lazard (1977) gave the proof for $d = -3$; it is similar to the one for $d \leq -15$, except that more subcases have to be considered. Incidentally, Lazard (1977) also showed analogous results for Z, the ring of (rational) integers, and for polynomial rings $K[x]$, K a field. The case $d = -4$, which was settled in Rolletschek (1986), is somewhat more complicated, see the discussion at the end of section 2. The remaining cases, namely $d = -7$, -8 or -11, are the main topic of this paper, and they will be dealt with in sections 3, 4. For $d = -7$ or -8 a new idea is needed, whereas for $d = -11$ a counterexample and a modified result are given. In section 5, finally, we discuss various questions that arise in connection with the proof.

We conclude this section by introducing some more notation (the same as in Kaltofen/Rolletschek (1985)), and by recalling some elementary facts about quadratic number fields, which can be found in Hasse (1964), or any other text on algebraic number theory.

We denote real and imaginary part of a complex number α by $\Re\,\alpha$ and $\Im\,\alpha$, its conjugate-complex by $\overline{\alpha}$, so that $\mathrm{N}(\alpha) = \alpha\overline{\alpha}$, and $|\alpha|$, the absolute value of α, equals $\sqrt{\mathrm{N}(\alpha)}$.

An algebraic number α is an *algebraic integer*, iff α is the root of some monic polynomial $x^n + a_{n-1}x^{n-1} + \cdots + a_0$ with integer coefficients $a_0, a_1, \ldots, a_{n-1} \in \mathrm{Z}$. Let $\mathbf{A}$ be the set of all algebraic integers. For D a squarefree integer the elements of $\mathbf{A} \cap \mathrm{Q}[\sqrt{D}\,]$ are the numbers of the form

$$\begin{cases} a + b\sqrt{D}, & a, b \in \mathrm{Z} \qquad\qquad\quad \text{if } D \equiv 2, 3 \bmod 4, \\ \frac{a+b\sqrt{D}}{2}, & a, b \in \mathrm{Z}, \quad a \equiv b \bmod 2 \quad \text{if } D \equiv 1 \bmod 4. \end{cases}$$

Alternatively, $\mathbf{A} \cap \mathrm{Q}[\sqrt{D}\,] = \{\, a + b\omega \mid a, b \in \mathrm{Z}\,\}$, where

$$\omega = \begin{cases} \sqrt{D} & \text{if } D \equiv 2, 3 \bmod 4, \\ \frac{1+\sqrt{D}}{2} & \text{if } D \equiv 1 \bmod 4. \end{cases}$$

This means that the set $1, \omega$ forms an *integral basis* for $\mathbf{A} \cap \mathrm{Q}[\sqrt{D}\,]$, and the number d used in the notation O_d for $\mathbf{A} \cap \mathrm{Q}[\sqrt{D}\,]$ is the *discriminant* of $\mathbf{A} \cap \mathrm{Q}[\sqrt{D}\,]$, defined by

$$d = \begin{vmatrix} 1 & \omega \\ 1 & \overline{\omega} \end{vmatrix} = \begin{cases} 4D & \text{if } D \equiv 2, 3 \bmod 4, \\ D & \text{if } D \equiv 1 \bmod 4. \end{cases}$$

From now on d will always have this meaning, whenever this symbol is used, for instance in the notation $\mathrm{Q}[\sqrt{d}\,]$.

In any ring R with 1 an element ϵ is called a *unit*, iff it has a multiplicative inverse, and the *associates* of an element $\alpha \in R$ are the elements of the form $\epsilon\alpha$ with ϵ a unit. The units in O_d, $d < 0$ are

$$\begin{cases} \pm 1, \frac{\pm 1 \pm \sqrt{-3}}{2} & \text{for } d = -3, \\ \pm 1, \pm i & \text{for } d = -4, \\ \pm 1 & \text{for } d < -4. \end{cases}$$

For $d > 0$ there are infinitely many units in O_d, but we will not need this fact.

A ring R is called Euclidean with respect to degree function ϕ, if ϕ is a function from R to $\mathbf{N}$ or some other well-ordered domain, and for all α, $\beta \in R$, $\beta \neq 0$, there exists a remainder $\alpha - \gamma\beta$ of α and β such that $\phi(\alpha - \gamma\beta) < \phi(\beta)$. This guarantees that

(EA) terminates for all inputs ρ_0, $\rho_1 \in R$, provided the quotients γ_1, γ_2, ... are chosen appropriately. It also implies that R is a unique factorization domain.

It is easy to show that O_d is Euclidean with respect to the norm for $d = -3, -4, -7, -8, -11$, but for no other $d < 0$. Stark (1967) showed that the only additional values of d for which O_d is a unique factorization domain are $d = -19, -43, -67$, and -163. As noted above, O_d is not Euclidean with respect to any degree function in these four cases.

Finally, whenever we consider a division chain $\Sigma = \langle \rho_0, \rho_1, \ldots \rangle$, the elements γ_i will always have the meaning specified in the clause (i) of the definition of division chains. Furthermore, we define

$$\omega_j = \frac{\rho_{j-1}}{\rho_j} \quad (1 \leq j < l(\Sigma)).$$

Then $1/\omega_{i+1} = (\rho_{i-1} - \gamma_i \rho_i)/\rho_i = \omega_i - \gamma_i$, hence

$$\omega_{i+1} = \frac{1}{\omega_i - \gamma_i}. \tag{1}$$

For division chains Σ', Σ'', ... we similarly use γ_i', ω_i', γ_i'', ω_i'' etc., without defining these numbers. By rest(Σ) we denote the sequence $\langle \rho_1, \rho_2, \ldots \rangle$ obtained from Σ by omitting the first element. It is again a division chain unless $l(\Sigma) = 0$.

2. The general proof method

Before giving the proof of the main result of this paper, we repeat its statement. Note that it suffices now to write $d \neq -11$ instead of $d \neq -11c^2$, due to the convention made after the definition of the discriminant.

THEOREM 1. *If $\Sigma = \langle \rho_0, \rho_1, \rho_2, \ldots \rangle$ is an arbitrary division chain, $\Sigma' = \langle \rho_0, \rho_1, \rho_2', \rho_3', \ldots \rangle$ a minimal remainder-division chain in $\mathbb{Q}[\sqrt{d}]$, $d < 0$, $d \neq -11$, where Σ and Σ' begin with the same two elements ρ_0, ρ_1, then $l(\Sigma') \leq l(\Sigma)$.*

Proof. The proofs for special cases and analogues of theorem 1, given in Lazard (1985), Rolletschek (1986) and Kaltofen/Rolletschek (1985) all proceed essentially by the same method. Our formulation will be very similar to that in (Rolletschek 1986). We apply induction on $l(\Sigma)$, so let Σ, Σ' be as in theorem 1, and assume that the assertion holds for every pair of division chains (Ω, Ω') in place of (Σ, Σ') such that $l(\Omega) < l(\Sigma)$. Without loss of generality we may make the following further assumptions:

(i) $\rho_1 = 1$; otherwise we may divide every element of Σ' and Σ by ρ_1.
(ii) $\gamma_1' = 0$, hence $\rho_2' = \rho_0$; otherwise we may replace ρ_0 by $\rho_0 - \gamma_1'\rho_1$, γ_1' by 0, γ_1 by $\gamma_1 - \gamma_1'$.

Assumption (ii) is equivalent to the assertion that 0 is one of the elements of O_d closest to ω_1, that is, that ρ_0 lies within the region R_0 shown in fig. 1. We make the convention that all regions defined are (topologically) closed unless stated otherwise. For instance, 1/2 is an element of R_0.

From Σ,Σ' we will construct division chains

$$\Sigma'' = \langle \rho_0, 1, \rho_2, \rho_3'', \rho_4'', \ldots \rangle, \qquad \Sigma''' = \langle \rho_0, 1, \rho_0, \rho_3''', \rho_4''', \ldots \rangle,$$

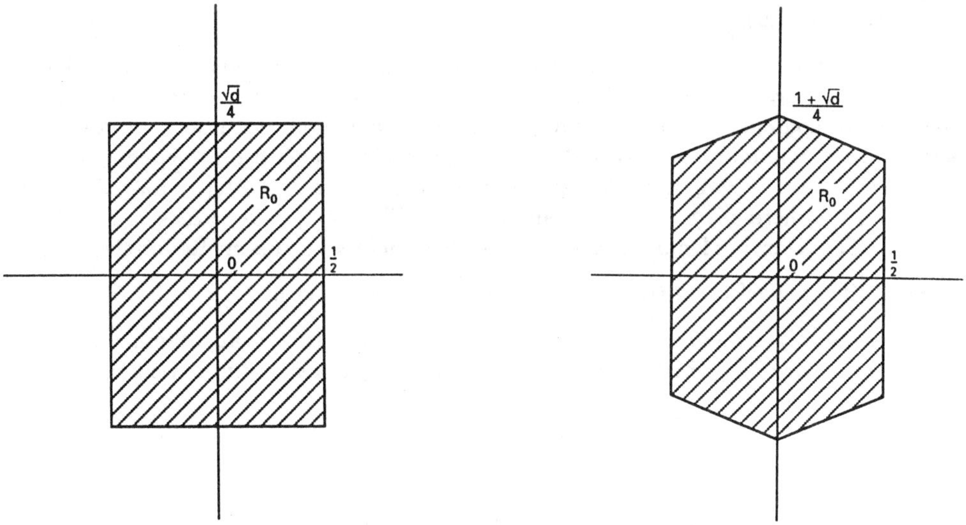

$d \equiv 0 \bmod 4$ $\qquad$ $d \equiv 1 \bmod 4$

Fig. 1

where the first three elements of Σ'' agree with those of Σ and the first three elements of Σ''' agree with those of Σ', such that

(a) $\mathrm{rest}(\Sigma'')$ is a minimal remainder division chain, and
(b) $l(\Sigma''') \leq l(\Sigma'')$.

If $l(\Sigma) = \infty$, then the assertion of the theorem holds trivially. Otherwise the induction hypothesis may be applied to $\mathrm{rest}(\Sigma'')$ and $\mathrm{rest}(\Sigma)$; it follows that $l(\mathrm{rest}(\Sigma'')) \leq l(\mathrm{rest}(\Sigma))$, hence $l(\Sigma''') \leq l(\Sigma'') \leq l(\Sigma)$. Since this implies $l(\mathrm{rest}(\Sigma''')) < l(\Sigma)$, we may now apply the induction hypothesis again to $\mathrm{rest}(\Sigma''')$ and $\mathrm{rest}(\Sigma')$ to conclude that $l(\Sigma') \leq l(\Sigma''')$. Combining, we obtain $l(\Sigma') \leq l(\Sigma)$, which is what has to be shown.

It remains to be shown that division chains Σ'', Σ''' of the desired form exist. Clearly, they depend only on ρ_0 and γ_1. In other words, we have to prove the following assertion, where we write again Σ, Σ' instead of Σ'', Σ''', since the original division chains Σ, Σ' no longer play a role:

(I) Let $\rho_0 \in \mathbf{Q}[\sqrt{d}]$, $\gamma_1 \in O_d$, $d < 0$, $d \neq -11$ be given, such that 0 is (one of) the minimal remainder-quotient(s) of ρ_0 and 1 (or equivalently, $\rho_0 \in R_0$). Then there exist division chains

$$\Sigma = \langle \rho_0, 1, \rho_2, \rho_3, \ldots \rangle, \qquad \Sigma' = \langle \rho_0, 1, \rho_0, \rho_3', \ldots \rangle,$$

such that

(a) γ_1, as defined at the end of section 1, has the given value,

(b) rest(Σ) is a minimal remainder-division chain,

(c) $l(\Sigma') \leq l(\Sigma)$.

We have stated (I) in the weakest form sufficient to prove theorem 1, so as to reduce the amount of work left. In particular, it will not be necessary to make Σ' a minimal remainder-division chain. Before proving (I), we need one more definition and a lemma. Given ξ, η, the minimal length of a division chain beginning with ξ, η depends only on $\omega = \xi/\eta$; denote it by #divs(ω). We say that two elements ω, $\omega' \in \mathrm{Q}[\sqrt{d}]$ are *d-related*, if $\omega' = \epsilon\omega + \alpha$, where ϵ is a unit of O_d and $\alpha \in O_d$. The equality $\omega' = \epsilon\omega + \alpha$ is called a *d-relation*. The following lemma is a slight modification of lemma 2 in (Rolletschek 1986).

LEMMA 1. *If ω and ω' are d-related, then #divs(ω) = #divs(ω').*

Proof. Assume $\omega' = \epsilon\omega + \alpha$. Let $\Sigma = \langle \omega, 1, \rho_2, \rho_3, \ldots \rangle$ be a division chain of minimal length for the given value of ω. Then $\Sigma' = \langle \omega', 1, \epsilon\rho_2, \rho_3, \epsilon\rho_4, \ldots \rangle$ is also a division chain, since

$$\omega' - \epsilon\rho_2 \in O_d$$

and

$$\begin{aligned} \rho_{i+1} &= \rho_{i-1} - (\epsilon^{-1}\gamma_i)(\epsilon\rho_i) && \text{for } i \text{ even,} \\ \epsilon\rho_{i+1} &= \epsilon\rho_{i-1} - (\epsilon\gamma_i)\rho_i && \text{for } i \text{ odd.} \end{aligned}$$

It follows that #divs(ω') $\leq$ #divs(ω). Since the notion of d-relation is symmetrical between ω and ω' ($\omega = \epsilon^{-1}\omega' - \epsilon^{-1}\alpha$), we also have #divs($\omega$) $\leq$ #divs(ω'). □

The proof of (I) proceeds by considering the various possible values of γ_1 separately, where $\rho_2 = \rho_0 - \gamma_1 \cdot 1$. Only finitely many values of γ_1 have to be considered, for if $|\gamma_1|$ is sufficiently large, then $|\rho_2| \geq 2$, which implies that 0 is a minimal remainder-quotient of 1 and ρ_2, so that we can choose

$$\begin{aligned} \Sigma &= \langle \rho_0, 1, \rho_2, 1, \rho_0, \rho_5, \ldots \rangle, \\ \Sigma' &= \langle \rho_0, 1, \rho_0, \rho_5, \ldots \rangle, \end{aligned} \tag{2}$$

where ρ_{i+1} is any minimal remainder of ρ_{i-1} and ρ_i for $i = 4, 5, \ldots$. Note that $\rho_2 - \rho_0 \in O_d$, hence ρ_0 is also a minimal remainder of ρ_2 and 1.

As an example of how the various cases can be dealt with, consider $d = -7$, $\gamma_1 = (1 + \sqrt{-7})/2$. The element $\rho_2 = \rho_0 - \gamma_1$ lies inside the region R_1 obtained by shifting R_0 to the left by 1/2 and down by $\sqrt{7}/2$, and ω_2 lies inside the region $R_1^{-1} = \{\alpha^{-1} \mid \alpha \in R_1\}$. One of the straight lines bounding R_1 is $l_1 = \{-1 + yi \mid y \in \mathrm{R}\}$. The inverse of a complex number $x + yi$ is $x/(x^2 + y^2) - yi/(x^2 + y^2)$, hence the inverses of the complex numbers on l_1 are the numbers $x + yi$ satisfying $x/(x^2 + y^2) = -1$, or $(x + 1/2)^2 + y^2 = 1/4$ (circle C in fig. 2). This is one of the circles bounding R_1^{-1}; the remaining bounds are determined analogously. Fig. 2 shows the regions R_1 and R_1^{-1}.

By elementary geometry it follows that either 0, or -1, or $(-1 + \sqrt{-7})/2$ is one of the elements of O_{-7} closest to ω_2 and may therefore be chosen as γ_2. Note that we need not consider the case $\gamma_2 = (1 + \sqrt{-7})/2$. True, in borderline cases ($\Re\,\omega_2 = 0$) $(1 + \sqrt{-7})/2$ is also a minimal remainder-quotient of 1 and ρ_2, but we can make the alternative choice

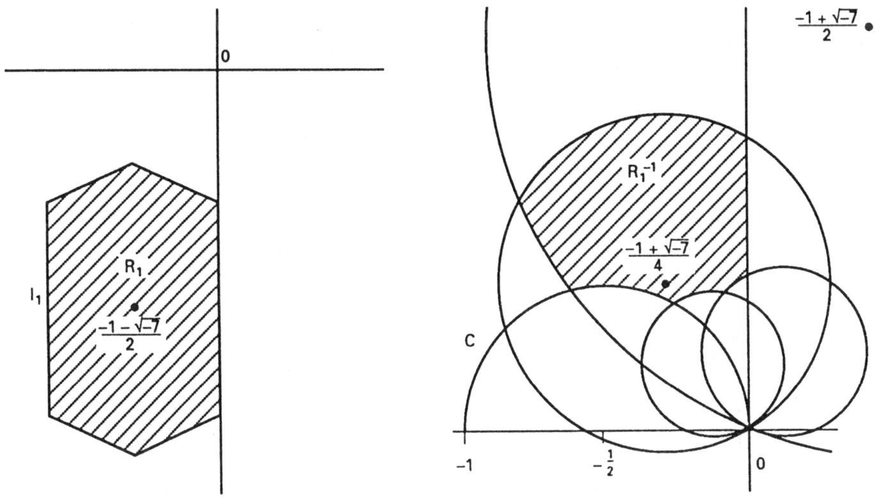

Fig. 2

$(-1+\sqrt{-7})/2$. The fact that for $i \geq 2$ γ_i may be chosen arbitrarily among the minimal remainders of ρ_{i-1} and ρ_i substantially reduces the number of cases to be considered.

The case where 0 is a possible choice has already been dealt with; here Σ and Σ' can be constructed as in (2). If $(-1+\sqrt{-7})/2$ is closest to ω_2, pick $\gamma_2 = (-1+\sqrt{-7})/2$, $\gamma_2' = (1-\sqrt{-7})/2$, so that

$$\Sigma = \left\langle \rho_0, 1, \rho_0 - \frac{1+\sqrt{-7}}{2}, \frac{1-\sqrt{-7}}{2}\rho_0 - 1, \rho_4, \ldots \right\rangle,$$

$$\Sigma' = \left\langle \rho_0, 1, \rho_0, \frac{-1+\sqrt{-7}}{2}\rho_0 + 1, \rho_4', \ldots \right\rangle.$$

Now either ρ_3 and ρ_3' are both 0, or we have

$$\omega_3 - \omega_3' = \frac{\rho_0 - \frac{1+\sqrt{-7}}{2} + \rho_0}{\frac{1-\sqrt{-7}}{2}\rho_0 - 1} = \frac{1+\sqrt{-7}}{2}.$$

The claim that we can make $l(\Sigma') \leq l(\Sigma)$ (in fact, $l(\Sigma') = l(\Sigma)$) is now guaranteed by lemma 1.

If, on the other hand, -1 is closest to ω_2, then ω_2 lies not only in R_1^{-1}, but also in the region $R_0 - 1$ obtained by shifting R_0 to the left by 1. Let R_2 denote the intersection of these two regions. From $\omega_2 \in R_2$ we can determine a region R_3 in which ω_3 must lie after choosing $\gamma_2 = -1$, based on the formula (1) at the end of section 1. Fig. 3 shows R_2 and R_3. Note that $1 \notin R_2$, hence $\rho_3 \neq 0$, hence ω_3 is always defined.

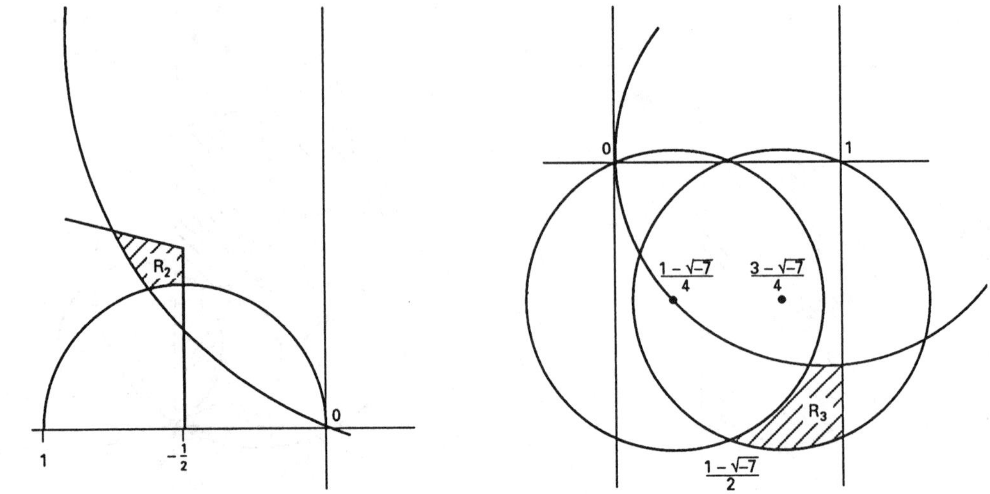

Fig. 3

This time $(1-\sqrt{-7})/2$ is always an element of O_{-7} closest to ω_3, hence put $\gamma_3 = (1-\sqrt{-7})/2$. Choose $\gamma_2' = (-1-\sqrt{-7})/2$. Similarly as in the previous paragraph we find that either $\rho_4 = \rho_3' = 0$, or $\omega_4 + \omega_3' = (1-\sqrt{-7})/2$, hence $l(\Sigma') = l(\Sigma) - 1$.

Many other cases—in fact, all cases for $d = -3$ and $d \leq -15$—can be settled in the same fashion. Table 1 provides a complete list of all cases that have to be considered, including those for $d = -11$. Each time the following information is given:

1. the sequence $\Gamma = \langle \gamma_1, \gamma_2, \ldots \rangle$, which characterizes the case;
2. the corresponding sequence $\Gamma' = \langle \gamma_1', \gamma_2', \ldots \rangle$;
3. a d-relation $\omega_j' = \epsilon\omega_i + \alpha$ between quotients ω_i, ω_j', $i \geq j$, which guarantees that $\#\text{divs}(\omega_i) = \#\text{divs}(\omega_j')$, that is, $l(\Sigma') \leq l(\Sigma)$. It should be interpreted in an extended sense, namely: either $\rho_i = \rho_j' = 0$ or $\omega_j' = \epsilon\omega_i + \alpha$.

In the cases for $d = -4$ i denotes the imaginary unit; no confusion will result with the use of the letter i as an index at other places. A question mark in columns 2 and 3 indicates that the case under consideration can *not* be dealt with in this simple way. A double question mark is used for the only case which cannot be dealt with at all—this is the case which causes the theorem to fail for $d = -11$, and counterexamples can be constructed from it.

We now provide a precise formulation of the assertion that table 1 is exhaustive, so that this claim can be justified. Given ρ_0, we call a sequence $\Gamma = \langle \gamma_1, \ldots, \gamma_{i-1} \rangle \in (O_d)^{i-1}$ an *admissible quotient-sequence* for ρ_0, if in the sequence $\Sigma = \langle \rho_0, 1, \rho_2, \ldots, \rho_i \rangle$ defined by $\rho_{k+1} = \rho_{k-1} - \gamma_k\rho_k$ for $k = 1, \ldots, i-1$, ρ_{k+1} is a minimal remainder of ρ_{k-1} and ρ_k for all $k \geq 2$. For $\omega_1 \in \mathrm{Q}[\sqrt{d}]$ and an arbitrary sequence Γ of elements of O_d define $\Omega(\omega_1, \Gamma)$

Γ	Γ'	d-relation
all $d<0$		
$\langle 1,-1\rangle$	$\langle 0\rangle$	$\omega_2' = -\omega_3 + 1$
$\langle 1,-2\rangle$	$\langle 0,2\rangle$	$\omega_3' = \omega_3 - 1$
$\langle 2,-1\rangle$	$\langle 0,1\rangle$	$\omega_3' = \omega_3 - 2$
$d=-3$		
$\langle 1, \frac{-3+\sqrt{-3}}{2}\rangle$	$\langle 0, \frac{3+\sqrt{-3}}{2}\rangle$	$\omega_3' = \frac{-1-\sqrt{-3}}{2}\omega_3 + \frac{-1+\sqrt{-3}}{2}$
$\langle \sqrt{-3}, \frac{1+\sqrt{-3}}{2}\rangle$	$\langle 0, \frac{1-\sqrt{-3}}{2}\rangle$	$\omega_3' = \frac{-1-\sqrt{-3}}{2}\omega_3 - \frac{3+\sqrt{-3}}{2}$
$\langle \sqrt{-3}, \frac{-1+\sqrt{-3}}{2}\rangle$	$\langle 0, \frac{-1-\sqrt{-3}}{2}\rangle$	$\omega_3' = \frac{-1+\sqrt{-3}}{2}\omega_3 + \frac{3-\sqrt{-3}}{2}$
$d=-4$		
$\langle 1,-1-i\rangle$	$\langle 0,1-i\rangle$	$\omega_3' = -\omega_3 - i$
$\langle 1,-2-i\rangle$	?	?
$\langle 1+i,-1\rangle$	$\langle 0,-i\rangle$	$\omega_3' = -\omega_3 + 1 - i$
$\langle 1+i,i\rangle$	$\langle 0,1\rangle$	$\omega_3' = -\omega_3 - 1 + i$
$\langle 1+i,-1+i\rangle$	$\langle 0,1-i\rangle$	$\omega_3' = \omega_3 - 1 - i$
$\langle 2+i,-1,2-i\rangle$	$\langle 0,1-i\rangle$	$\omega_3' = -\omega_4 - 1 - i$
$d=-7$		
$\langle 1, \frac{-1+\sqrt{-7}}{2}, \frac{-1+\sqrt{-7}}{2}\rangle$	$\langle 0, \frac{1+\sqrt{-7}}{2}\rangle$	$\omega_3' = -\omega_4 + \frac{1+\sqrt{-7}}{2}$
$\langle 1, \frac{-1+\sqrt{-7}}{2}, \frac{-3+\sqrt{-7}}{2}\rangle$	$\langle 0, \frac{1+\sqrt{-7}}{2}, \frac{3+\sqrt{-7}}{2}\rangle$	$\omega_4' = \omega_4 - 1$
$\langle 1, \frac{-3+\sqrt{-7}}{2}\rangle$	?	?
$\langle \frac{1+\sqrt{-7}}{2}, -1, \frac{1-\sqrt{-7}}{2}\rangle$	$\langle 0, \frac{-1-\sqrt{-7}}{2}\rangle$	$\omega_3' = -\omega_4 + \frac{1-\sqrt{-7}}{2}$
$\langle \frac{1+\sqrt{-7}}{2}, \frac{-1+\sqrt{-7}}{2}\rangle$	$\langle 0, \frac{1-\sqrt{-7}}{2}\rangle$	$\omega_3' = \omega_3 - \frac{1+\sqrt{-7}}{2}$
$\langle \frac{3+\sqrt{-7}}{2}, -1, \frac{3-\sqrt{-7}}{2}\rangle$	$\langle 0, \frac{1-\sqrt{-7}}{2}\rangle$	$\omega_3' = -\omega_4 - \frac{1+\sqrt{-7}}{2}$
$d=-8$		
$\langle 1,-1+\sqrt{-2}\rangle$	?	?
$\langle 1,-2+\sqrt{-2}\rangle$	?	?
$\langle \sqrt{-2},\sqrt{-2}\rangle$	$\langle 0,-\sqrt{-2}\rangle$	$\omega_3' = \omega_3 - \sqrt{-2}$
$\langle \sqrt{-2},1+\sqrt{-2}\rangle$	?	?
$\langle 1+\sqrt{-2},\sqrt{-2},-1+\sqrt{-2}\rangle$	$\langle 0,1-\sqrt{-2},-\sqrt{-2}\rangle$	$\omega_4' = \omega_4 - (1+\sqrt{-2})$
$\langle 1+\sqrt{-2},-1,1-\sqrt{-2}\rangle$	$\langle 0,-\sqrt{-2}\rangle$	$\omega_3' = -\omega_4 - \sqrt{-2}$
$\langle 1+\sqrt{-2},-1,2-\sqrt{-2}\rangle$	?	?
$\left\{ \langle 1+\sqrt{-2},-1+\sqrt{-2}, 1+\sqrt{-2}\rangle \right.$	$\langle 0,1-\sqrt{-2},-1-\sqrt{-2}\rangle$	$\omega_4' = \omega_4 + 1 - \sqrt{-2}$
$\left\{ \langle 2+\sqrt{-2},-1,2-\sqrt{-2}, -1-\sqrt{-2}\rangle \right.$	$\langle 0,1-\sqrt{-2},-1-\sqrt{-2}\rangle$	$\omega_4' = -\omega_5 + 1 - \sqrt{-2}$

Table 1

Γ	Γ'	d-relation
$d=-11$		
$\langle 1, \frac{-1+\sqrt{-11}}{2}\rangle$	?	?
$\langle 1, \frac{-3+\sqrt{-11}}{2}\rangle$	?	?
$\langle \frac{1+\sqrt{-11}}{2}, -1, \frac{1-\sqrt{-11}}{2}, \frac{1-\sqrt{-11}}{2}\rangle$	$\langle 0, \frac{-1-\sqrt{-11}}{2}, \frac{1-\sqrt{-11}}{2}\rangle$	$\omega_4' = -\omega_5 - \frac{1+\sqrt{-11}}{2}$
$\langle \frac{1+\sqrt{-11}}{2}, \frac{-1+\sqrt{-11}}{2}, \frac{1+\sqrt{-11}}{2}\rangle$	$\langle 0, \frac{1-\sqrt{-11}}{2}, \frac{-1-\sqrt{-11}}{2}\rangle$	$\omega_4' = \omega_4 + \frac{1-\sqrt{-11}}{2}$
$\langle \frac{1+\sqrt{-11}}{2}, \frac{-1+\sqrt{-11}}{2}, \frac{-1+\sqrt{-11}}{2}\rangle$	??	??
$\langle \frac{3+\sqrt{-11}}{2}, -1, \frac{3-\sqrt{-11}}{2}, \frac{-1-\sqrt{-11}}{2}\rangle$	$\langle 0, \frac{1-\sqrt{-11}}{2}, \frac{-1-\sqrt{-11}}{2}\rangle$	$\omega_4' = -\omega_5 + \frac{1-\sqrt{-11}}{2}$
$\langle \frac{3+\sqrt{-11}}{2}, -1, \frac{3-\sqrt{-11}}{2}, \frac{1-\sqrt{-11}}{2}\rangle$	$\left\{ \langle 0, \frac{1-\sqrt{-11}}{2}, \frac{-1-\sqrt{-11}}{2}, \frac{-1-\sqrt{-11}}{2}\rangle \right.$	$\omega_5' = \omega_5 + 1$

Table 1 (cont.)

recursively by

$$\Omega(\omega_1, \Gamma) = \omega_1 \quad \text{if } \Gamma \text{ is the empty sequence,}$$
$$\Omega(\omega_1, \langle \gamma_1, \ldots, \gamma_{i+1}\rangle) = \frac{1}{\Omega(\omega_1, \langle \gamma_1, \ldots, \gamma_i\rangle) - \gamma_{i+1}}.$$

Thus for a division chain $\Sigma = \langle \rho_0, \rho_1, \ldots\rangle$, it follows from the equality (1) at the end of section 1 that $\omega_i = \Omega(\omega_1, \langle \gamma_1, \ldots, \gamma_{i-1}\rangle)$. We say that a sequence $\Gamma' = \langle \gamma_1', \ldots, \gamma_{j-1}'\rangle$ ***improves*** (***strictly improves***) $\Gamma = \langle \gamma_1, \ldots, \gamma_{i-1}\rangle$, if $j \leq i$ $(j < i)$, and there is a d-relation between $\Omega(\omega_1, \Gamma)$ and $\Omega(\omega_1, \Gamma')$ for every value of (or: independently of) ω_1. For $\alpha \in O_d$ Γ is (***strictly***) ***α-improvable***, if there exists a sequence Γ' beginning with α which (strictly) improves Γ. It follows easily from lemma 1 that the following assertion implies (I):

(II) For every $\rho_0 \in Q[\sqrt{d}]$, $d < 0$, $d \neq -11$ such that $\rho_0 \in R_0$, and every $\gamma_1 \in O_d$, there exists an admissible quotient-sequence Γ for ρ_0 beginning with γ_1 which is 0-improvable.

It turns out that not all values of γ_1 have to be considered: after dealing with some value of γ_1, say γ, we may omit the cases $\gamma_1 = \epsilon\gamma$ or $\gamma_1 = \overline{\epsilon\gamma}$ with ϵ a unit. This is a consequence of the following lemma 2. For a sequence $\Gamma = \langle \gamma_1, \gamma_2, \ldots, \gamma_i\rangle$ and a unit ϵ define

$$\Gamma_\epsilon = \langle \epsilon\gamma_1, \epsilon^{-1}\gamma_2, \ldots, \epsilon^{(-1)^{i+1}}\gamma_i\rangle,$$
$$\overline{\Gamma} = \langle \overline{\gamma_1}, \ldots, \overline{\gamma_i}\rangle.$$

LEMMA 2.

(a) *If Γ is (strictly) 0-improvable, then so are Γ_ϵ and $\overline{\Gamma_\epsilon}$ for every unit ϵ.*
(b) *Every sequence of the form $\langle \gamma_1, 0\rangle$, $\gamma_1 \in O_d$, is strictly 0-improvable.*

Proof.

(a) Assume $\Gamma' = \langle\gamma_1', \ldots, \gamma_j'\rangle$ (strictly) improves $\Gamma = \langle\gamma_1, \ldots, \gamma_i\rangle$. Let ω_1 be arbitrary. Induction on i shows that $\Omega(\epsilon^{-1}\omega_1, \Gamma) = \epsilon^{(-1)^{i+1}}\Omega(\omega_1, \Gamma_\epsilon)$, and similarly $\Omega(\epsilon^{-1}\omega_1, \Gamma') = \epsilon^{(-1)^{j+1}}\Omega(\omega_1, \Gamma_\epsilon')$. By assumption, $\Omega(\epsilon^{-1}\omega_1, \Gamma)$ and $\Omega(\epsilon^{-1}\omega_1, \Gamma')$ are d-related. Since d-relatedness is transitive, it follows that $\Omega(\omega_1, \Gamma_\epsilon)$ and $\Omega(\omega_1, \Gamma_\epsilon')$ are also d-related, hence Γ_ϵ' (stricly) improves Γ_ϵ. Also, if Γ' begins with 0, then so does Γ_ϵ'. This completes the proof that Γ_ϵ is (strictly) 0-improvable. The automorphism of complex conjugation can be applied to Γ_ϵ, Γ_ϵ' and the resulting d-relation, hence $\overline{\Gamma_\epsilon}$ is also (strictly) 0-improvable.

(b) This was already shown immediately after lemma 1. □

We can now show that table 1 is complete in the following sense:

LEMMA 3. *For every $\rho_0 \in R_0$ and every $\gamma_1 \in O_d$ there exists an admissible quotient-sequence $\Gamma = \langle\gamma_1, \ldots, \gamma_i\rangle$ for ρ_0 such that either $\gamma_2 = 0$ or Γ_ϵ or $\overline{\Gamma_\epsilon}$ is considered in table 1, that is, occurs in the leftmost column of table 1.*

Proof. We note first that for every value of γ_1, either (i) $\langle\gamma_1, 0\rangle$ is an admissible quotient-sequence for every $\rho_0 \in R_0$, or (ii) for some unit ϵ sequences Γ beginning with $\epsilon\gamma_1$ or $\overline{\epsilon\gamma_1}$ occur in the leftmost column of table 1. For instance, for $d = -4$ the alternative (ii) holds if γ_1 has one of the values $0, \pm 1, \pm i, \pm 1 \pm i, \pm 2, \pm 2i, \pm 2 \pm i, \pm 1 \pm 2i$, whereas in all other cases $|\gamma_1| \geq 2\sqrt{2}$, so that for every $\rho_0 \in R_0$, $|\rho_0 - \gamma_1 \cdot 1| \geq 2\sqrt{2} - \sqrt{2}/2 > 2$, $|\omega_2| < 1/2$, hence $\langle\gamma_1, 0\rangle$ is indeed an admissible quotient-sequence for ρ_0.

Second, for every value of γ_1 considered in table 1, and for every $\rho_0 \in R_0$, there exists an admissible quotient-sequence Γ for ρ_0 beginning with γ_1, such that either Γ or $\overline{\Gamma}$ occurs in the leftmost column of table 1. This was shown in the sample case above for $d = -7$, $\gamma_1 = (1 + \sqrt{-7})/2$: in this case, either $\langle(1 + \sqrt{-7})/2, -1, (1 - \sqrt{-7})/2\rangle$ or $\langle(1 + \sqrt{-7})/2, (-1 + \sqrt{-7})/2\rangle$ is admissible for ρ_0 for every $\rho_0 \in R_0$. In all other cases the proof is similar. Instances where $\overline{\Gamma}$ rather than Γ appears in table 1 can of course occur only for γ_1 real, since for $\gamma \neq \overline{\gamma}$ only one of these values of γ_1 is considered in table 1; in fact, such instances occur only for $\gamma_1 = 1$, for instance, if $d = -3$, $\gamma_1 = 1$, then for every $\rho_0 \in R_0$ either $\langle 1, (-3 + \sqrt{-3})/2\rangle$ or $\langle 1, (-3 - \sqrt{-3})/2\rangle$ is an admissible quotient-sequence for ρ_0.

Finally, consider arbitrary elements ρ_0, γ_1, as specified in the statement of the lemma. We assume that the leftmost column of table 1 contains sequences Γ beginning with $\epsilon\gamma_1$; the proof for the case where such sequences begin with $\overline{\epsilon\gamma_1}$ is similar. As noted in the previous paragraph, there exists an admissible quotient-sequence Γ' for $\rho_0' = \epsilon\rho_0$ beginning with $\epsilon\gamma_1$ such that Γ' or $\overline{\Gamma'}$ occurs in table 1. To complete the proof of lemma 3, we have to show that $\Gamma = \Gamma'_{\epsilon^{-1}}$ is an admissible quotient-sequence for γ_1. Put $\Gamma = \langle\gamma_1, \gamma_2, \ldots\rangle$, $\Gamma' = \langle\gamma_1', \gamma_2', \ldots\rangle$ $(= \langle\epsilon\gamma_1, \epsilon^{-1}\gamma_2, \ldots\rangle)$, and define sequences $\Sigma = \langle\rho_0, 1, \rho_2, \ldots\rangle$, $\Sigma' = \langle\rho_0', 1, \rho_2', \ldots\rangle$ by $\rho_{i+1} = \rho_{i-1} - \gamma_i\rho_i$, $\rho_{i+1}' = \rho_{i-1}' - \gamma_i'\rho_i'$ for $i = 1, 2, \ldots$. Then induction on k shows that

$$\rho_k' = \begin{cases} \epsilon\rho_k & \text{for } k \text{ even,} \\ \rho_k & \text{for } k \text{ odd.} \end{cases}$$

The fact that Γ' is an admissible quotient-sequence for ρ_0' means that for all $k \geq 2$, ρ_{k+1}' is a minimal remainder of ρ_{k-1}' and ρ_k'; what is left to show is that ρ_{k+1} is also a minimal remainder of ρ_{k-1} and ρ_k. If k is even, then the remainders of ρ_{k-1} and ρ_k coincide with those ρ'_{k-1} and ρ'_k, while for k odd the remainders of ρ_{k-1} and ρ_k are the numbers of the form $\epsilon^{-1}\rho$, where ρ is a remainder of ρ'_{k-1} and ρ'_k. In both cases the assertion follows. This completes the proof of lemma 3. □

As a special case of lemma 3 we note that for $d \le -15$, only the first three sequences Γ of table 1 have to be considered, which apply to all $d < 0$.

We now turn to those cases which cannot be settled like the sample case above, as indicated by the question marks. Here the problem is that there exists no (constant) bound on the smallest j, k for which a d-relation between ω_j and ω_k exists. Therefore these cases split into an *infinite* number of subcases.

Consider first $d = -4$, $\gamma_1 = 1$, $\gamma_2 = -2 - i$. If ρ_0 is very close to $(1 + (\sqrt{2} - 1)i)/2$, then $\gamma_j = 2(-1)^{j+1} - 2i$ for $j = 3, 4, \ldots$, for a number of steps which may be arbitrarily large. A d-relation is finally obtained between ω_{l+1} and either ω'_l or ω'_{l+1}, where l is the smallest j for which $\gamma_j \neq 2(-1)^{j+1} - 2i$. The proof of this relation is carried out in (Rolletschek 1986) in the following fashion: since the quotients γ_3, γ_4, ..., and also the quotients $\gamma'_3, \gamma'_4, \ldots$ of the sequence Σ' to be constructed follow a periodic pattern, ρ_l, ρ_{l+1}, ρ'_l, ρ'_{l+1} can be expressed in closed form by solving a linear recurrence relation. The rest is then a straightforward calculation.

This method does not work, however, for some of the remaining cases for $d = -7$, -8, since the quotients γ_j are no longer necessarily periodic. Therefore a new approach is needed. We introduce it in the next section.

3. Cases with unbounded look-ahead

Among the cases left open in section 2, the only one for $d = -7$ is $\gamma_1 = 1$, $\gamma_2 = (-3 + \sqrt{-7})/2$. As before we determine the regions where ω_1, ω_2 can lie. It turns out that despite the fact that there is no bound for the number of steps we have to look ahead in the sequences Σ and Σ', only a finite number of regions has to be considered altogether. First, $\gamma_1 = 1$, $\gamma_2 = (-3+\sqrt{-7})/2$ implies that $\omega_2 \in R_4$, $\omega_3 \in R_5$, where R_4, R_5 are shown in fig. 4.

If for any i $\omega_i \in R_5$, then the element $\gamma_i \in O_d$ closest to ω_i may be $(-1 + \sqrt{-7})/2$, $(1 + \sqrt{-7})/2$, $(3 + \sqrt{-7})/2$, $\sqrt{-7}$ or $1 + \sqrt{-7}$. If $\gamma_i = \sqrt{-7}$, then ω_{i+1} lies again in R_5, while if $\gamma_i = 1 + \sqrt{-7}$, then ω_{i+1} lies in the region R_6 obtained from R_5 by reflection on the imaginary axis. The situation is of course analogous for $\omega_j \in R_6$: in this case γ_j may be $(-3 + \sqrt{-7})/2$, $(-1 + \sqrt{-7})/2$, $(1 + \sqrt{-7})/2$, $\sqrt{-7}$ or $-1 + \sqrt{-7}$, and we have

$$\omega_{j+1} \in \begin{cases} R_5, & \text{if } \gamma_i = -1 + \sqrt{-7}, \\ R_6, & \text{if } \gamma_i = \sqrt{-7}. \end{cases}$$

As soon as for some i $\omega_i \in R_5$ and $\gamma_i \in \{(-1 + \sqrt{-7})/2, (1 + \sqrt{-7})/2, (3 + \sqrt{-7})/2\}$, or $\omega_i \in R_6$ and $\gamma_i \in \{(-3 + \sqrt{-7})/2, (-1 + \sqrt{-7})/2, (1 + \sqrt{-7})/2\}$, a d-relation between ω_{i+1} and either ω'_i or ω'_{i+1} can be established, as will be shown below. That such an i exists follows from the following two facts:

(i) The division chain Σ is finite, since O_{-7} is Euclidean with respect to the norm.

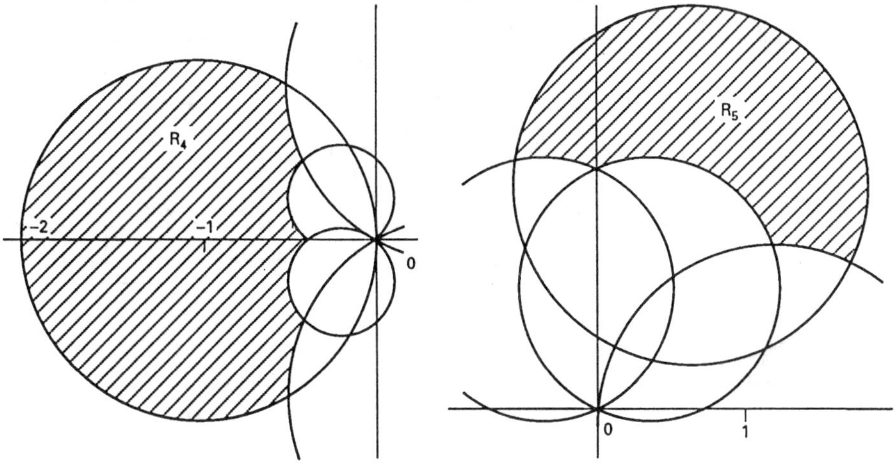

Fig. 4

(ii) $\sqrt{-7}, 1+\sqrt{-7} \notin R_5$ and $\sqrt{-7}, -1+\sqrt{-7} \notin R_6$, hence if $\omega_i \in R_5$ and $\gamma_i \in \{\sqrt{-7}, 1+\sqrt{-7}\}$, or $\omega_i \in R_6$ and $\gamma_i \in \{\sqrt{-7}, -1+\sqrt{-7}\}$, then $\rho_{i+1} \neq 0$.

Let l be the smallest $i \geq 3$ for which $\gamma_i \notin \{-1+\sqrt{-7}, \sqrt{-7}, 1+\sqrt{-7}\}$. The crucial observation, which allows to complete the proof for the current case, as well as for all the remaining cases for $d=-8$, is the following. If we choose $\gamma_i' = -\overline{\gamma_i}$ for $i = 2, 3, \ldots, l-1$, so that $\Re\,\gamma_i' = -\Re\,\gamma_i$, $\Im\,\gamma_i' = \Im\,\gamma_i$, then a certain relation e_k between ω_j and ω_j' is satisfied whenever $\omega_j \in R_k$, namely:

$$e_4\colon\ \omega_2\omega_2' + \omega_2' - \omega_2 = 0$$

$$e_5\colon\ \omega_j\omega_j' - \frac{1+\sqrt{-7}}{2}\omega_j' + \frac{1-\sqrt{-7}}{2}\omega_j - 1 = 0$$

$$e_6\colon\ \omega_j\omega_j' + \frac{1-\sqrt{-7}}{2}\omega_j' - \frac{1+\sqrt{-7}}{2}\omega_j - 1 = 0.$$

In the equation e_4 we wrote ω_2 rather than ω_j, since the region R_4 applies only for ω_2. True, these relations do not by themselves imply $\#\mathrm{divs}(\omega_j') \leq \#\mathrm{divs}(\omega_j)$ directly. However, they can be proven in a straightforward way by induction, and after they have been established, a d-relation between ω_{l+1} and ω_l' or ω_{l+1}' follows in a similar way. As an induction basis we have

$$\frac{1}{\omega_2} = \rho_0 - 1 = \frac{1}{\omega_2'} - 1,$$

which simplifies to the equality e_4. From the equality (1) at the end of section 1 we get $\omega_i = 1/\omega_{i+1} + \gamma_i$, $\omega_i' = 1/\omega_{i+1}' + \gamma_i'$, in particular, $\omega_2 = 1/\omega_3 + (-3+\sqrt{-7})/2$,

region for ω_l	γ_l	d-relation
R_5	$\frac{-1+\sqrt{-7}}{2}$	e_7: $\omega'_{l+1} = \omega_{l+1} - 1$
R_5	$\frac{1+\sqrt{-7}}{2}$	e_9: $\omega'_l = -\omega_{l+1} + \frac{-1+\sqrt{-7}}{2}$
R_5	$\frac{3+\sqrt{-7}}{2}$	e_8: $\omega'_{l+1} = \omega_{l+1} + 1$
R_6	$\frac{-3+\sqrt{-7}}{2}$	e_7: $\omega'_{l+1} = \omega_{l+1} - 1$
R_6	$\frac{-1+\sqrt{-7}}{2}$	e_{10}: $\omega'_l = -\omega_{l+1} + \frac{1+\sqrt{-7}}{2}$
R_6	$\frac{1+\sqrt{-7}}{2}$	e_8: $\omega'_{l+1} = \omega_{l+1} + 1$

Table 2

$\omega'_2 = 1/\omega'_3 + (3+\sqrt{-7})/2$. We substitute these expressions for ω_2, ω'_2 in e_4; this yields the equation e_5 for $j = 3$ after simplification. Now assume e_5 holds for some j, where $\omega_j \in R_5$. Suppose $\gamma_j = \sqrt{-7}$. Recall that $\omega_{j+1} \in R_5$ in this case. By substituting $1/\omega_{j+1} + \sqrt{-7}$ for ω_j, $1/\omega'_{j+1} + \sqrt{-7}$ for ω'_j, we obtain e_5 with $j+1$ in place of j. In the remaining cases the induction step is carried out similarly. Finally, we obtain by the same method the d-relations between ω_{l+1} and ω'_l or ω'_{l+1} given in table 2.

In those cases where a d-relation between ω_{l+1} and ω'_{l+1} rather than ω'_l is obtained, we have to pick $\gamma'_i = -\overline{\gamma_i}$ not only for $i = 2, 3, \ldots, l-1$, but also for $i = l$. These d-relations imply that $l(\Sigma') \leq l(\Sigma)$, as in previous cases. This completes the proof of the theorem for $d = -7$.

The crucial pieces of information for this proof are contained in the following:

(i) the definitions of the regions R_i,
(ii) the definition of the γ'_i's in dependence of the γ_i's,
(iii) the equations e_i, and
(iv) the transition diagram shown in fig. 5.

This transition diagram has to be interpreted as follows: the label of every node is a condition C_i for some ω_k, ω'_k. Whenever the node is circled only once, then C_i means: $\omega_k \in R_i$ and the equation e_i is satisfied. When it is circled twice, then it means that e_i is satisfied; moreover, e_i is a d-relation. No region R_i for ω_k is specified in this case; in fact, such a region would have to be unbounded in many instances, and ω_k might also be undefined ($\rho_k = 0$). Consider a node circled only once and labelled C_i. Then the labels of the arrows out of this node are the values of γ_k that have to be considered under the assumption that ω_k satisfies C_i. These arrows point to nodes whose labels indicate which conditions are satisfied by ω_{k+1} for each value of γ_k. The label of the "initial" arrow pointing to the node labelled C_4 is the value of γ_1, hence C_4 is satisfied by ω_2 and ω'_2.

For the remaining cases the proof proceeds along the same lines, and we provide only the information (i)–(iv) above. A table shows for each j the following:

(a) a list the possible values of γ_k under the condition $\omega_k \in R_j$; these values are given under the heading "possible γ_k".
(b) The corresponding equalities e_j.

When C_j is an "exit condition", circled twice in the transition diagram, then R_j is

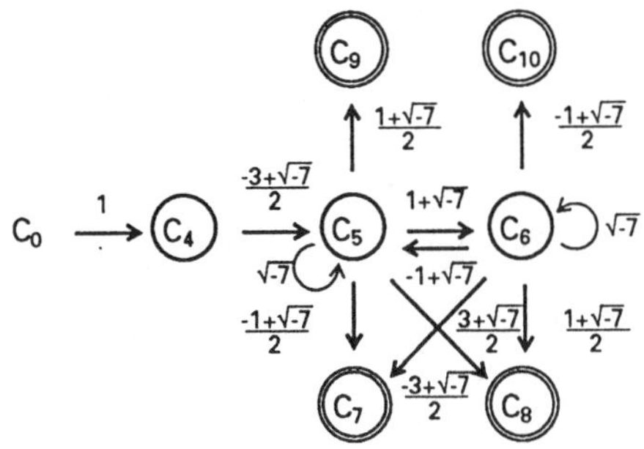

Fig. 5

unspecified, hence the possible values of γ_k are omitted. They are also omitted, when only one value of γ_k falls under the case under consideration. For instance, in the next case considered, R_{11} is the set of possible values of ω_2 after putting $\gamma_1 = 1$. Now only the value $\gamma_2 = -2-i$ is considered, since all other values were already dealt with in section 2. Hence the possible values of γ_2 for $\omega_2 \in R_{11}$ are omitted in the table. All transition diagrams have to be interpreted as explaned in the previous paragraph.

Case $d = -4$, $\gamma_1 = 1$, $\gamma_2 = -2 - i$: this case was already dealt with in (Rolletschek 1986), as noted at the end of section 2. Using the method of the current section, we obtain a different proof, which requires somewhat less calculation, since no recurrence relation has to be solved. Analogously to the case above, the index l denotes the smallest k for which γ_k "leads to an exit condition", that is, ω_{l+1} satisfies one of the equalities e_{14}–e_{18}. Again we have to choose $\gamma'_k = -\overline{\gamma_k}$ for $k = 2, \ldots, l-1$, and also for $k = l$, if the exit condition obtained is between ω_{l+1} and ω'_{l+1}. The relevant information is given in table 3 and in fig. 6, 7. Region R_{13} is not drawn; it is obtained from R_{12} by reflection on the imaginary axis. The remaining cases all deal with discriminant $d = -8$. The greatest variety of regions, namely R_{19}–R_{24}, occurs for the case $\gamma_1 = 1$, $\gamma_2 = -1 + \sqrt{-2}$ or $-2 + \sqrt{-2}$; this is the only case where the corresponding transition diagram contains cycles of length 3. Fig. 8, 9 show regions R_{19}, R_{20}, R_{21} and R_{23}, whereas R_{22} and R_{24} are the mirror images of R_{21} and R_{23} about the imaginary axis. All other information is contained in table 4 and fig. 10. We note that after choosing $\gamma_1 = 1$, $\gamma_2 = -1 + \sqrt{-2}$, ω_3 can actually be shown to lie in a region slightly smaller that R_{20}, but the same is not true for ω_k, if $\omega_{k-1} \in R_{20}$ and $\gamma_{k-1} = 2\sqrt{-2}$. For this reason R_{20} cannot be replaced by a smaller region. The same situation occurs in other cases, too. Of course, it does not effect the proof, if some region R_i properly contains the set of all possible values of ω_k for certain choices of $\gamma_1, \ldots, \gamma_{k-1}$, while only $\omega_k \in R_i$ is assumed.

The last two cases are similar to $d = -4$, $\gamma_1 = 1$, $\gamma_2 = -2 - i$, since $\gamma_3, \gamma_4, \ldots, \gamma_{l-1}$ are periodic. Hence the proof could once again be carried out by solving a linear recurrence relation. The next case considered is $\gamma_1 = \sqrt{-2}$, $\gamma_2 = \pm 1 + \sqrt{-2}$, see table 5 and fig. 11, 12. The region R_{32} is the mirror image of R_{31}, its picture is omitted. We have to choose $\gamma'_k = \overline{\gamma_k}$ for $k = 2, 3, \ldots, l$. Now comes the last case: $d = -8$, $\gamma_1 = 1 + \sqrt{-2}$, $\gamma_2 = -1$, see table 6 and fig. 13, 14. As can be seen from the transition diagram, we have $\gamma_k = 2(-1)^{k+1} - \sqrt{-2}$ for $k = 3, \ldots, l-1$, $\gamma_l = (-1)^{l+1} - \sqrt{-2}$. This case differs from others in so far as all equations e_j except e_{34} relate ω_k with ω'_{k-1} rather than ω'_k. The transition diagram is based on the following choice of the γ'_k's: if $l = 3$, then $\gamma'_2 = -\sqrt{-2}$, while if $l > 3$, then $\gamma'_2 = -1 - \sqrt{-2}$, $\gamma'_k = \gamma_k$ for $k = 3, \ldots, l-2$, and $\gamma'_{l-1} = (-1)^l - \sqrt{-2}$. In this way we get $l(\Sigma') = l(\Sigma) - 1$. Again one of the regions, namely R_{36}, is not drawn, since it is the mirror image of R_{35} about the imaginary axis. Also, no new region R_{37} is defined for this case, since the region associated with C_{37} is again R_{35}. Thus C_{37} is the condition that $\omega_k \in R_{35}$ and e_{37} holds. □

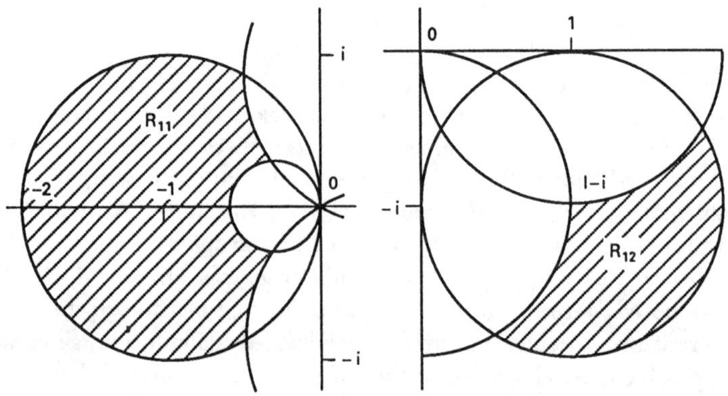

Fig. 6: regions R_{11}, R_{12}

j	possible γ_k	equality e_j
11		$\omega_2\omega'_2 + \omega'_2 - \omega_2 = 0$
12	$1-i, 2-i, 1-2i, 2-2i$	$\omega_k\omega'_k - (1-i)\omega'_k + (1+i)\omega_k - 1 = 0$
13	$-1-i, -2-i, -1-2i, -2-2i$	$\omega_k\omega'_k + (1+i)\omega'_k - (1-i)\omega_k - 1 = 0$
14		$\omega'_l = -\omega_{l+1} - 1 - i$
15		$\omega'_{l+1} = \omega_{l+1} + 1$
16		$\omega'_{l+1} = -\omega_{l+1} - i$
17		$\omega'_l = -\omega_{l+1} + 1 - i$
18		$\omega'_{l+1} = \omega_{l+1} - 1$

Table 3

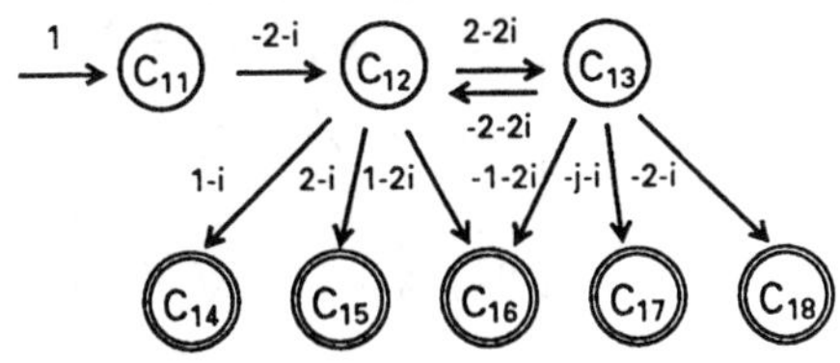

Fig. 7: Transition diagram for the case $d = -4$, $\gamma_1 = 1$, $\gamma_2 = -2 - i$

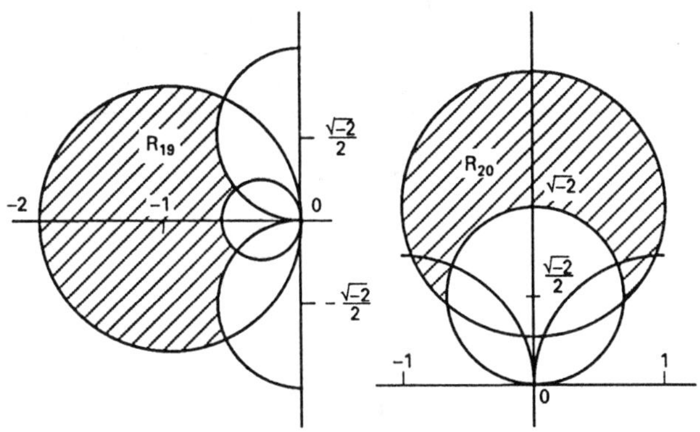

Fig. 8: regions R_{19} and R_{20}

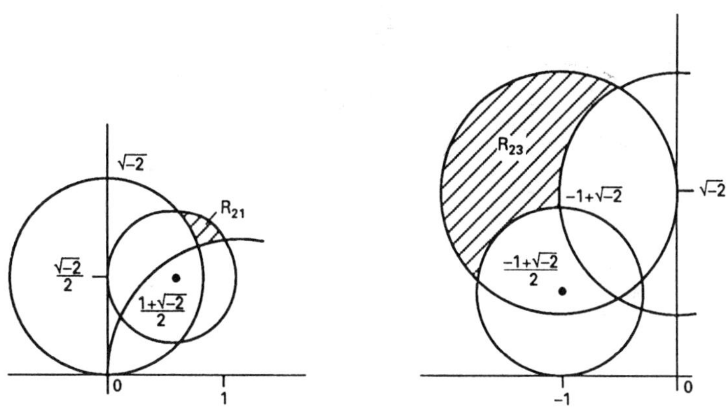

Fig. 9: regions R_{21} and R_{23}

j	possible γ_k	equality e_j
19	$-1, -2, -1 \pm \sqrt{-2}, -2 \pm \sqrt{-2}$	$\omega_2\omega_2' + \omega_2' - \omega_2 = 0$
20	$\begin{cases} -1+\sqrt{-2}, \sqrt{-2}, 1+\sqrt{-2}, \\ -1+2\sqrt{-2}, 2\sqrt{-2}, 1+2\sqrt{-2} \end{cases}$	$\omega_2\omega_2' - \sqrt{-2}\,\omega_2' - \sqrt{-2}\,\omega_2 - 1 = 0$
21	$1+\sqrt{-2}$	$2\omega_2\omega_2' - (1+\sqrt{-2})\omega_2' + (1-\sqrt{-2})\omega_2 - 1 = 0$
22	$-1+\sqrt{-2}$	$2\omega_2\omega_2' + (1-\sqrt{-2})\omega_2' - (1+\sqrt{-2})\omega_2 - 1 = 0$
23	$\begin{cases} -2+\sqrt{-2}, -1+\sqrt{-2}, \\ -2+2\sqrt{-2}, -1+2\sqrt{-2} \end{cases}$	$\omega_2\omega_2' + (1-\sqrt{-2})\omega_2' - (1+\sqrt{-2})\omega_2 - 2 = 0$
24	$\begin{cases} 1+\sqrt{-2}, 2+\sqrt{-2}, \\ 1+2\sqrt{-2}, 2+2\sqrt{-2} \end{cases}$	$\omega_2\omega_2' - (1+\sqrt{-2})\omega_2' + (1-\sqrt{-2})\omega_2 - 2 = 0$
25		$\omega_{l+1}' = \omega_{l+1} - 1$
26		$\omega_{l+1}' = \omega_{l+1} + 1$
27		$\omega_l' = -\omega_{l+1} + \sqrt{-2}$
28		$\omega_l' = -\omega_{l+1} + 1 + \sqrt{-2}$
29		$\omega_l' = -\omega_{l+1} - 1 + \sqrt{-2}$

Table 4

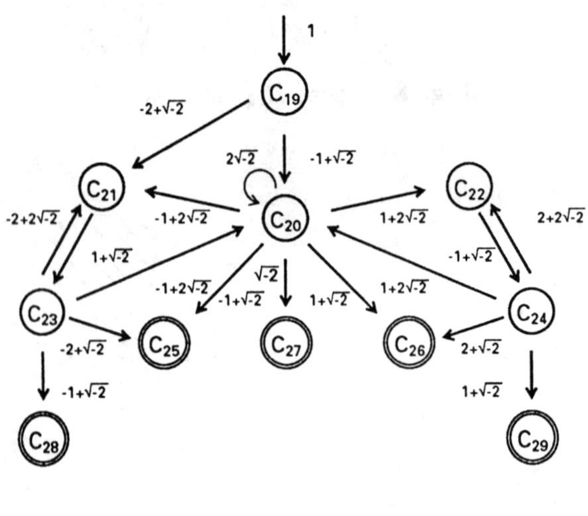

Fig. 10: transition diagram for $d = -8$, $\gamma_1 = 1, \gamma_2 \in \{-1+\sqrt{-2}, -2+\sqrt{-2}\}$

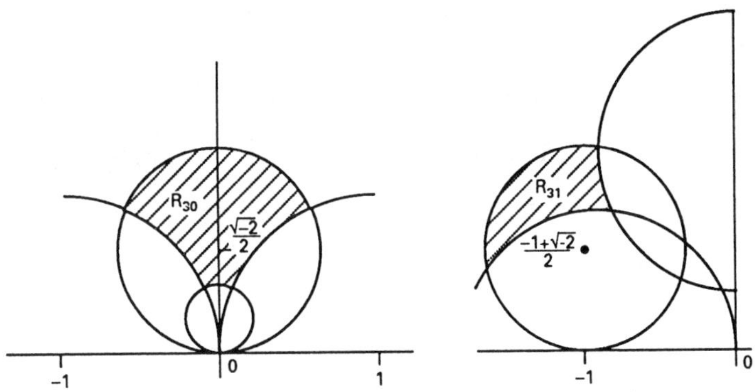

Fig. 11: regions R_{30} and R_{31}

j	possible γ_k	equality e_j
30		$\sqrt{-2}\,\omega_2\omega_2' + \omega_2' - \omega_2 = 0$
31	$-1+\sqrt{-2}, -2+\sqrt{-2}$	$\sqrt{-2}\,\omega_k\omega_k' + (1+\sqrt{-2})\omega_k' - (1-\sqrt{-2})\omega_k + \sqrt{-2} = 0$
32	$1+\sqrt{-2}, 2+\sqrt{-2}$	$\sqrt{-2}\,\omega_k\omega_k' + (1-\sqrt{-2})\omega_k' - (1+\sqrt{-2})\omega_k + \sqrt{-2} = 0$
33		$\omega_{l+1}' = \omega_{l+1} - \sqrt{-2}$

Table 5

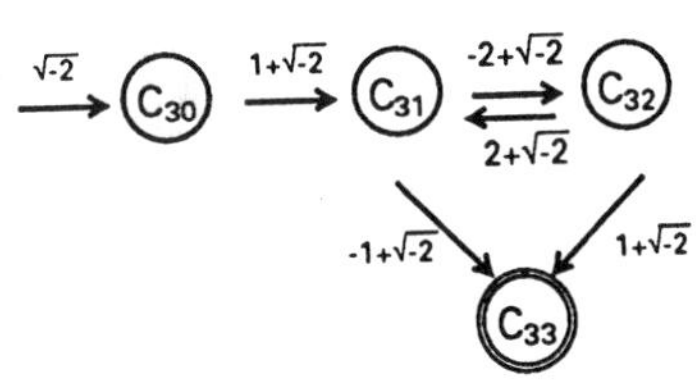

Fig. 12: transition diagram for $d = -8$, $\gamma_1 = \sqrt{-2}$, $\gamma_2 = 1+\sqrt{-2}$

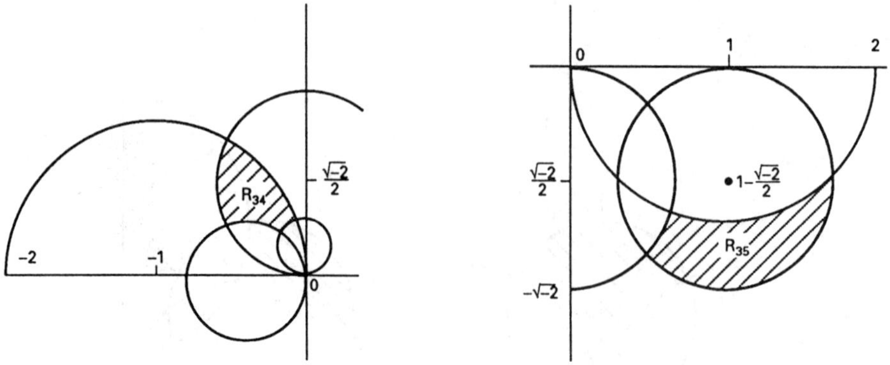

Fig. 13: regions R_{34} and R_{35}

j	possible γ_k	equality e_j
34		$(1+\sqrt{-2})\omega_2\omega_2' + \omega_2' - \omega_2 = 0$
35	$1-\sqrt{-2}, 2-\sqrt{-2}$	$\sqrt{-2}\,\omega_k\omega_{k-1}' - (1+\sqrt{-2})\omega_{k-1}' - \omega_k + 1 = 0$
36	$-1-\sqrt{-2}, -2-\sqrt{-2}$	$\begin{cases} \sqrt{-2}\,\omega_k\omega_{k-1}' + (-1+\sqrt{-2})\omega_{k-1}' \\ \quad -(1+\sqrt{-2})\omega_k - \sqrt{-2} = 0 \end{cases}$
37	$1-\sqrt{-2}, 2-\sqrt{-2}$	$\begin{cases} \sqrt{-2}\,\omega_k\omega_{k-1}' - (1+\sqrt{-2})\omega_{k-1}' \\ \quad +(-1+\sqrt{-2})\omega_k - \sqrt{-2} = 0 \end{cases}$
38		$\omega_{l+1}' = -\omega_l - \sqrt{-2}$

Table 6

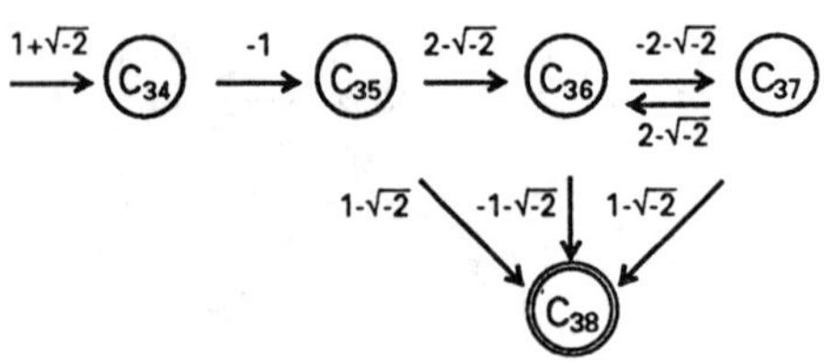

Fig. 14: transition diagram for $d=-8$, $\gamma_1 = 1+\sqrt{-2}$, $\gamma_2 = -1$

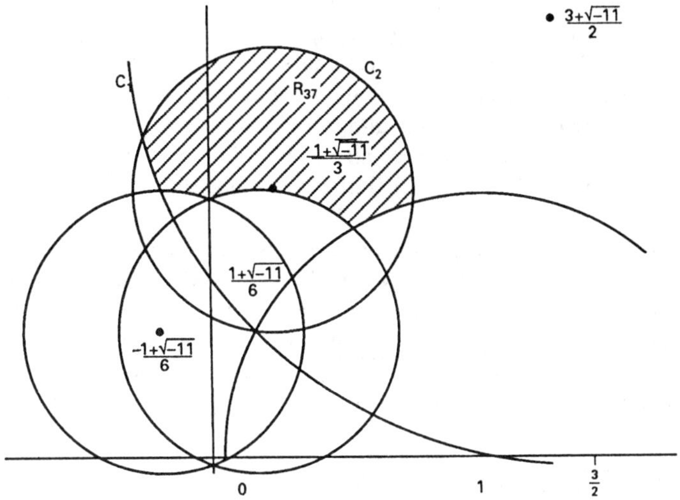

Fig. 15: region R_{37}

4. The case d = −11

We claimed in section 2 that there is one case for $d = -11$ where the proof method applied for all other cases does not work, and where the theorem actually fails, namely: $\gamma_1 = (1+\sqrt{-11})/2$, $\gamma_2 = (-1+\sqrt{-11})/2$, $\gamma_3 = (-1+\sqrt{-11})/2$. In order to investigate this case more closely, look at the region R_{37} containing all values ω_3 can have after choosing $\gamma_1 = (1+\sqrt{-11})/2$, $\gamma_2 = (-1+\sqrt{-11})/2$, assuming that $\omega_1 \in R_0$ and that $(-1+\sqrt{-11})/2$ is one of the elements of O_d closest to ω_2, as in other cases considered in the proof. R_{37} is shown in fig. 15.

One of the values of ω_3 which requires to choose $\gamma_3 = (-1+\sqrt{-11})/2$ is the one at the intersection of the circles labelled C_1 and C_2, namely $(-1+2\sqrt{-11})/5$. It occurs for $\omega_1 = -3/\sqrt{-11}$, for instance, $\rho_0 = 6$, $\rho_1 = -2\sqrt{-11}$. For these values of ρ_0, ρ_1 it turns out that if we choose $\gamma_1 = 0$, we have to carry out one more division than if we choose $\gamma_1 = (1+\sqrt{-11})/2$. The resulting division chains are

$$\Sigma = \langle 6, -2\sqrt{-11}, -5+\sqrt{-11}, 3+\sqrt{-11}, 2, 0\rangle$$
$$\Sigma' = \langle 6, -2\sqrt{-11}, 6, -3+\sqrt{-11}, -1-\sqrt{-11}, 2, 0\rangle.$$

We have thus obtained the first counterexample, showing that theorem 1 does not carry over to the domain O_{-11}.

The phrase 'the number of divisions is minimized by choosing $\gamma_1 = \ldots$' can be formalized by the following definition: an element $\gamma \in O_d$ is an *m-quotient* of ξ and η, iff $\#\mathrm{divs}(\eta/(\xi-\gamma\eta)) = \#\mathrm{divs}(\xi/\eta)-1$. In this case there does indeed exist a shortest division chain beginning with ξ, η whose third term is $\xi - \gamma\eta$. We also call such an element γ an *m-approximation* of $\omega = \xi/\eta$. With this definition theorem 1 could have been stated in the following equivalent form: in every imaginary quadratic number ring O_d, $d < 0$, $d \neq -11$, every minimal remainder-quotient of two elements ξ, η is also an m-quotient of ξ, η.

In the example just considered, we have not yet shown that γ_i is an m-approximation of ω_i for $i = 2, 3, 4$, and that γ_i' is an m-approximation of ω_i' for $i = 2, \ldots, 5$; this will follow from theorem 2 below. However, all γ_i' are minimal remainder-quotients of ρ_{i-1} and ρ_i, and it follows from $l(\Sigma') > l(\Sigma)$ that at least one of them is not an m-quotient of ρ_{i-1} and ρ_i.

Note that 0 and $(1+\sqrt{-11})/2$ are both minimal remainder-quotients of ρ_0 and ρ_1. There are also other counterexamples, where 0 is the *only* minimal remainder-quotient of ρ_0 and ρ_1, and where $\#\mathrm{divs}(\rho_1/\rho_0) > \#\mathrm{divs}(\rho_1/(\rho_0 - \rho_1(1+\sqrt{-11})/2))$. They are obtained by picking values of ω_1 different from, but close to $-3/\sqrt{-11}$, for instance, $\omega_1 = -3/\sqrt{-11} + (1-\sqrt{-11})/77 = (1+20\sqrt{-11})/77$, where the choices $\gamma_1 = (1+\sqrt{-11})/2$ and $\gamma_1' = 0$ result in the remainder sequences

$$\Sigma = \Big\langle 1+20\sqrt{-11}, 77, \frac{-75-37\sqrt{-11}}{2}, \frac{-87+19\sqrt{-11}}{2}, -7+8\sqrt{-11},$$
$$11+\sqrt{-11}, -7+2\sqrt{-11}, \frac{-7-3\sqrt{-11}}{2}, \frac{-1-\sqrt{-11}}{2}, 1, 0\Big\rangle,$$
$$\Sigma' = \Big\langle 1+20\sqrt{-11}, 77, 1+20\sqrt{-11}, \frac{-67-19\sqrt{-11}}{2}, 3-11\sqrt{-11}, \frac{51-5\sqrt{-11}}{2},$$
$$4+3\sqrt{-11}, \frac{-15+3\sqrt{-11}}{2}, \frac{-1-3\sqrt{-11}}{2}, \frac{1+\sqrt{-11}}{2}, 1, 0\Big\rangle.$$

In this example, as in the one considered before, both $(1+\sqrt{-11})/2$ and $(-1+\sqrt{-11})/2$ are m-quotients of ρ_0 and ρ_1.

It is natural to ask for a rule for obtaining an m-approximation for any given value of ω_i. (Although the practical importance of such a rule for calculating gcd's is very small, see the discussion in section 5.) However, the situation seems to be fairly complicated, and a complete answer has not yet been obtained. In the rest of this section, a partial answer will be given. The first step is to determine the set of all values of ω_1 which lead to the case $\gamma_1 = (1+\sqrt{-11})/2$, $\gamma_2 = (-1+\sqrt{-11})/2$, $\gamma_3 = (-1+\sqrt{-11})/2$, that is, for which, $(-1+\sqrt{-11})/2$ is the only element of O_{-11} closest to ω_2, and also the only element closest to ω_3, after choosing $\gamma_1 = (1+\sqrt{-11})/2$, $\gamma_2 = (-1+\sqrt{-11})/2$. This is readily accomplished; we denote it by R_{38}', see fig. 16. Two of its bounds are straight lines bounding R_0, whereas the third is the circle with center $(-1+\sqrt{-11})/2$ and radius 1. This circle is not included in R_{38}', that is, R_{38} is not bounded. If ω_1 lies on this circle, then γ_3 could also be chosen as $(1+\sqrt{-11})/2$ instead of $(-1+\sqrt{-11})/2$, since ω_3 would lie on the imaginary axis.

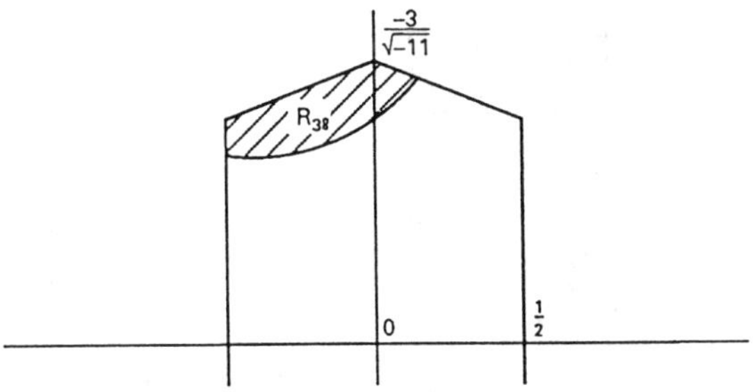

Fig. 16: region R'_{38}

Surprisingly, the major part of R'_{38} consists of values ω_1 such that $\Re\,\omega_1 < 0$. It is plausible, and will follow from theorem 2, that whenever $\Re\,\omega_1 < 0$ and $(1+\sqrt{-11})/2$ is an m-approximation of ω_1, then $(-1+\sqrt{-11})/2$ is also an m-approximation of ω_1. When it comes to deciding between $\gamma_1 = 0$ and $\gamma_1 = (1+\sqrt{-11})/2$, then the values ω_1 such that $\Re\,\omega_1 < 0$ may be ignored, since for such ω_1 the decision should be made between $\gamma_1 = 0$ and $\gamma_1 = (-1+\sqrt{-11})/2$. Correspondingly, we define $R_{38} = R'_{38} \cap \{\omega_1 \mid \Re\,\omega_1 \geq 0\}$. The discussion given so far suggests the following conjecture: suppose $\omega_1 \in R_0$, $\Re\,\omega_1 \geq 0$, $\Im\omega_1 \geq 0$. If $\omega_1 \notin R_{38}$, then $\gamma_1 = 0$ is an m-approximation of ω_1. If $\omega_1 \in R_{38}$, then either 0, or $(1+\sqrt{-11})/2$, or both are m-approximations of ω_1. The following theorem generalizes this conjecture to arbitrary values of ω_1.

THEOREM 2. *Let $\omega \in \mathrm{Q}[\sqrt{-11}]$ be given. Let κ, λ be elements of O_{-11} satisfying the equalities and inequalities*

$$\Re\,\kappa \leq \Re\,\omega \leq \Re\,\kappa + \frac{1}{2} = \Re\,\lambda,$$

$$|\Im\,\kappa - \Im\,\lambda| = \frac{\sqrt{-11}}{2},$$

$$\Im\,\kappa \leq \Im\,\omega \leq \Im\,\lambda \text{ or } \Im\,\lambda \leq \Im\,\omega \leq \Im\,\kappa.$$

(Equivalently, κ and λ may be characterized by stating that $\kappa, \lambda \in O_{-11}$, $\lambda - \kappa = (1 \pm \sqrt{-11})/2$, and κ and λ are opposite corners of a rectangle containing ω with sides parallel to the real and imaginary axis. They are uniquely determined, unless $\Re\,\omega = k/2$ or $\Im\,\omega = k\sqrt{-11}/2$ for some rational integer k.) If $|\kappa - \omega| \leq |\lambda - \omega|$ and $|(\lambda - 1) - \omega| \geq 1$, then κ is an m-approximation for ω. If $|\lambda - \omega| \leq |\kappa - \omega|$ and $|(\kappa + 1) - \omega| \geq 1$, then λ is an m-approximation for ω. In the remaining cases $|\kappa - \omega| \leq |\lambda - \omega|$ and $|(\lambda - 1) - \omega| < 1$, or $|\lambda - \omega| \leq |\kappa - \omega|$ and $|(\kappa + 1) - \omega| < 1$) at least one of the elements κ or λ is an m-approximation for ω.

Proof. The method is the same as for theorem 1. Similarly to assertion (I) for theorem 1, it suffices to show

(III) Let $\rho_0 \in \mathrm{Q}[\sqrt{-11}] \cap R_0$, $\gamma_1 \in O_{-11}$ be given. Then there exist division chains

$$\Sigma = \langle \rho_0, 1, \rho_2, \rho_3, \ldots \rangle, \qquad \Sigma' = \langle \rho_0, 1, \rho'_2, \rho'_3, \ldots \rangle,$$

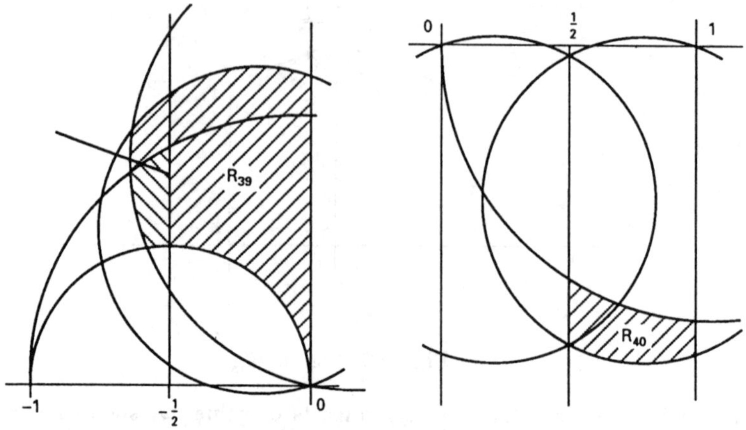

Fig. 17: regions R_{39} and R_{40}

such that

(a) γ_1, as defined for sequences Σ, has the specified value;
(b) For every $i \geq 2$ γ_i is an m-quotient of ρ_{i-1} and ρ_i;
(c)

$$\gamma_1' = \begin{cases} 0 \text{ or } \frac{1+\sqrt{-11}}{2} & \text{if } \Re\,\omega_1 \geq 0 \text{ and } |\omega_1 - \frac{-1+\sqrt{-11}}{2}| < 1, \\ 0 \text{ or } \frac{-1+\sqrt{-11}}{2} & \text{if } \Re\,\omega_1 \leq 0 \text{ and } |\omega_1 - \frac{1+\sqrt{-11}}{2}| < 1, \\ 0 \text{ or } \frac{-1-\sqrt{-11}}{2} & \text{if } \Re\,\omega_1 \leq 0 \text{ and } |\omega_1 - \frac{1-\sqrt{-11}}{2}| < 1, \\ 0 \text{ or } \frac{1-\sqrt{-11}}{2} & \text{if } \Re\,\omega_1 \geq 0 \text{ and } |\omega_1 - \frac{-1-\sqrt{-11}}{2}| < 1, \\ 0 & \text{otherwise;} \end{cases}$$

(d) $l(\Sigma') \leq l(\Sigma)$.

For any given values ρ_0, γ_1, (III) is equivalent to $\#\text{divs}(1/(\omega_1 - \gamma_1')) \leq \#\text{divs}(1/(\omega_1 - \gamma_1))$, with γ_1' as specified in (c). When checking condition (b), we are allowed to use theorem 2 as a criterion for m-quotients (induction hypothesis). In the proof of theorem 1 the induction hypothesis was used only to reduce the theorem to (I). Here, however, it would be difficult to formulate (III) in complete analogy to (I), since theorem 2 only partially characterizes m-quotients. If for some $i \geq 2$ $\omega_i \in R_{38}$, then two different sequences Σ must be constructed, one with $\gamma_i = 0$ and one with $\gamma_i = (1 + \sqrt{-11})/2$, since we do not know which is an m-approximation for ω_i. On the other hand, if $\omega_i = 1/2 + ai$ for some small a, then it suffices to choose either $\gamma_i = 0$ or $\gamma_i = 1$, as in similar cases in the proof of (I), since both 0 and 1 are m-approximations of ω_i.

Despite this ambiguity, it turns out that the set of cases considered in table 1 is complete (in the sense explaned in section 2). For instance, suppose $\gamma_1 = (1 + \sqrt{-11})/2$. Then ω_2 lies in the region R_{39} shown in fig. 17. It is easy to see that according to theorem 2, only 0, -1, or $(-1 + \sqrt{-11})/2$ can be an m-approximation of ω_2; it can be -1, if ω_2 lies in the portion of R_{39} shaded \\\\\\\\. Then, if $\gamma_2 = -1$, then $\omega_3 \in R_{40}$, where R_{40} is a region also shown in fig. 17. R_{40} contains no points ω_3 sufficiently close to $1/2+3/\sqrt{-11}$, so that 0 or 1 could be an m-approximation for ω_3; such points would have to lie inside the circles with radius 1 and centers 0 and 1, but the second of these circles in excluded from R_{40}. Similarly, we find that only $(1-\sqrt{-11})/2$ can be an m-approximation for ω_4, which settles the case $\gamma_1 = (1+\sqrt{-11})/2$, $\gamma_2 = -1$.

The only cases left are those marked ? and ?? in table 1, namely (x) $\gamma_1 = 1$ and (y) $\gamma_1 = (1+\sqrt{-11})/2$, $\gamma_2 = (-1+\sqrt{-11})/2$, $\gamma_3 = (-1+\sqrt{-11})/2$. As for case (x), we will show below that the conditions in (III) can be satisfied with $\gamma_1' = 0$, even if the condition $\rho_0 \in R_0$ is replaced by $\Re\,\rho_0 \leq 1/2$. Suppose for the moment that this has been done already. As noted earlier, case (y) can only occur if $\omega_1 \in R_{38}'$, as a straightforward calculation shows. Assume $\omega_1 \in R_{38}'$. If $\Re\,\omega_1 \geq 0$, then $\omega_1 \in R_{38}$, and we have the case where according to theorem 2 $(1+\sqrt{-11})/2$ may be an m-approximation for ω_1. In this case the conditions in (III) are trivially satisfied by putting $\gamma_1' = (1+\sqrt{-11})/2$. If $\Re\,\omega_1 < 0$, then $\Re(\omega_1 + (1-\sqrt{-11})/2) < 1/2$, and by our assumption about case (x) we have

$$\begin{aligned}\#\mathrm{divs}\left(\frac{1}{\omega_1-\gamma_1}\right) &= \#\mathrm{divs}\left(\frac{1}{(\omega_1+(1-\sqrt{-11})/2)-1}\right)\\ &\geq \#\mathrm{divs}\left(\frac{1}{(\omega_1+(1-\sqrt{-11})/2)-0}\right)\\ &= \#\mathrm{divs}\left(\frac{1}{\omega_1-(-1+\sqrt{-11})/2}\right).\end{aligned}$$

It follows that a sequence Σ' satisfying $l(\Sigma') \leq l(\Sigma)$ can be constructed by choosing $\gamma_1' = (-1+\sqrt{-11})/2$. If $|\omega_1 - (1+\sqrt{-11})/2| < 1$, then this value of γ_1' also satisfies condition (c) of (III), and we are done. Otherwise we still have to show that a division chain Σ' with $\gamma_1' = (-1+\sqrt{-11})/2$ can be replaced by one with $\gamma_1' = 0$ without increasing its length, that is, we have to show

$$\#\mathrm{divs}\left(\frac{1}{\omega_1-(-1+\sqrt{-11})/2}\right) \geq \#\mathrm{divs}\left(\frac{1}{\omega_1-0}\right),$$

which amounts to proving (III) for an instance with $\gamma_1 = (-1+\sqrt{-11})/2$ and $\Re\,\omega_1 < 0$. This is symmetrical to the case $\gamma_1 = (1+\sqrt{-11})/2$, $\Re\,\omega_1 > 0$, which has already been dealt with.

It remains to consider the case (x), that is, $\gamma_1 = 1$, with the condition $\rho_0 \in R_0$ replaced by $\Re\rho_0 \leq 1/2$. Here we have to apply the method of section 3. We need seven more regions R_{41}–R_{47}, of which R_{41}, R_{42}, R_{44} and R_{46} are shown in fig. 18 and 19, whereas for $i = 43$, 45 and 47 R_i is the mirror image of R_{i-1} about the imaginary axis. We have to choose $\gamma_k' = \overline{-\gamma_k}$ for $k = 2, \ldots,$ until an exit condition is reached. The remaining information is provided in table 7 and in the transition diagram, fig. 20.

Similarly as in various cases in section 3, we note that if the transition diagram indicates that $\omega_k \in R_i$ for certain values of $\gamma_1, \ldots, \gamma_{k-1}$, then the set of ω_k which are possible in this case may in fact be a proper subset of R_i. For instance, if $\omega_{k-1} \in R_{42}$, $\gamma_{k-1} = (1+\sqrt{-11})/2$, then it can be concluded that ω_k must lie within a region considerably smaller than R_{46}. However, we are forced to include additional values in R_{46},

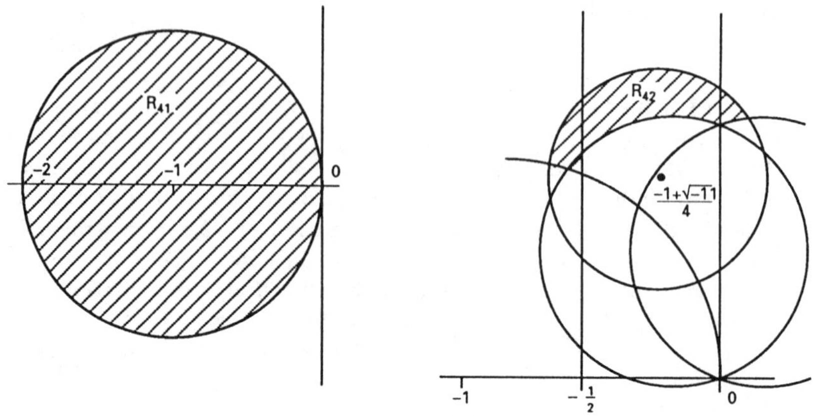

Fig. 18: regions R_{41} and R_{42}

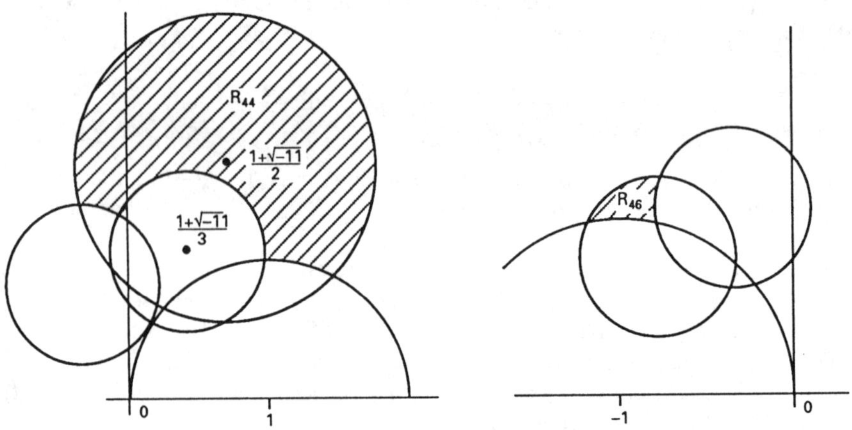

Fig. 19: regions R_{44} and R_{46}

j	possible γ_k	equality e_j
41	$-2,-1,0,\frac{-3\pm\sqrt{-11}}{2},\frac{-1\pm\sqrt{-11}}{2}$	$\omega_1\omega_1'+\omega_1'-\omega_1=0$
42	$\frac{-1+\sqrt{-11}}{2},\frac{1+\sqrt{-11}}{2}$	$2\omega_k\omega_k'+\frac{1-\sqrt{-11}}{2}\omega_k'-\frac{1+\sqrt{-11}}{2}\omega_k-1=0$
43	$\frac{-1+\sqrt{-11}}{2},\frac{1+\sqrt{-11}}{2}$	$2\omega_k\omega_k'-\frac{1+\sqrt{-11}}{2}\omega_k'+\frac{1-\sqrt{-11}}{2}\omega_k-1=0$
44	$\begin{cases}\frac{-1+\sqrt{-11}}{2},\frac{1+\sqrt{-11}}{2},\frac{3+\sqrt{-11}}{2},\\ \sqrt{-11},1+\sqrt{-11}\end{cases}$	$\omega_k\omega_k'-\frac{1+\sqrt{-11}}{2}\omega_k'+\frac{1-\sqrt{-11}}{2}\omega_k-2=0$
45	$\begin{cases}\frac{-3+\sqrt{-11}}{2},\frac{-1+\sqrt{-11}}{2},\frac{1+\sqrt{-11}}{2},\\ -1+\sqrt{-11},\sqrt{-11}\end{cases}$	$\omega_k\omega_k'+\frac{1-\sqrt{-11}}{2}\omega_k'-\frac{1+\sqrt{-11}}{2}\omega_k-2=0$
46	$\frac{-3+\sqrt{-11}}{2},\frac{-1+\sqrt{-11}}{2}$	$2\omega_k\omega_k'+\frac{3-\sqrt{-11}}{2}\omega_k'-\frac{3+\sqrt{-11}}{2}\omega_k-2=0$
47	$\frac{1+\sqrt{-11}}{2},\frac{3+\sqrt{-11}}{2}$	$2\omega_k\omega_k'-\frac{3+\sqrt{-11}}{2}\omega_k'+\frac{3-\sqrt{-11}}{2}\omega_k-2=0$
48		$\omega_{l+1}'=\omega_{l+1}-1$
49		$\omega_{l+1}'=\omega_{l+1}+1$
50		$\omega_l'=-\omega_{l+1}-\frac{1-\sqrt{-11}}{2}$
51		$\omega_l'=-\omega_{l+1}+\frac{1+\sqrt{-11}}{2}$

Table 7

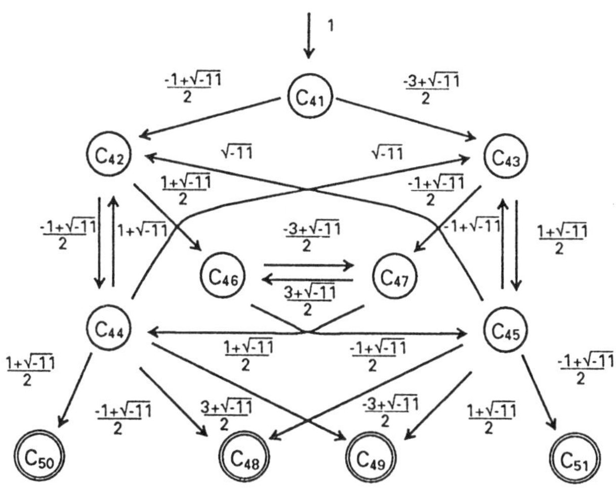

Fig. 20: transition diagram for $d=-11$, $\gamma_1=1$

otherwise $\omega_{k-1} \in R_{47}$ and $\gamma_{k-1} = (3+\sqrt{-11})/2$ would not imply $\omega_k \in R_{46}$. Again, this has no effect on the proof of theorem 2, which is now complete. □

A number of test computations have been carried out in an attempt to determine, for which $\omega_1 \in R_{38}$ it is better to choose $\gamma_1 = 0$, and for which it is better to choose $\gamma_1 = (1+\sqrt{-11})/2$. Define

$$\omega_1 \in R_{38} \text{ is of type } \begin{Bmatrix} A \\ B \\ C \end{Bmatrix} \text{ iff } \#\text{divs}(\frac{1}{\omega_1 - (1+\sqrt{-11})/2}) \begin{Bmatrix} > \\ = \\ < \end{Bmatrix} \#\text{divs}(\frac{1}{\omega_1 - 0}).$$

Thus if ω_1 is of type A (C), then one has to choose $\gamma_1 = 0$ ($\gamma_1 = (1+\sqrt{-11})/2$), in order to minimize the number of divisions, whereas for ω_1 of type B it makes no difference. What follows are the results of these test computations. Note that theorem 2 does provide a method for finding an m-approximation for every $\omega \in \mathrm{Q}[\sqrt{-11}]$, since we may choose ρ_0, ρ_1 such that $\rho_0/\rho_1 = \omega_1$, and systematically try all division chains $\Sigma = \langle \rho_0, \rho_1, \rho_2, \ldots \rangle$ for which, according to theorem 2, it may be the case that γ_i is an m-approximation of ω_i for all i. This primitive method, though of exponential complexity, is practical enough, since the "borderline cases" are quite rare.

Let $\alpha = (\sqrt{5} - 2 + \sqrt{-11})/4$; this is one of the two points where the line $x + (1/4)i$ ($x \in \mathrm{R}$) and the circle $|(x+yi) - (-1+\sqrt{-11})/2| = 1$ intersect; hence α lies at the boundary of R_{38}. Within small neighborhoods of α ($|\omega_1 - \alpha| < 10^{-6}$) values of ω_1 of each of the types A, B, C were found, and it seems very likely that this is true for every neighborhood of α, and that α is the only point in $\overline{R_{38}}$ with this property. ($\overline{R_{38}}$ denotes the (topological) closure of R_{38}.) Take α as a reference point, and put $\omega_1 = \alpha + x + yi$ ($x, y \in \mathrm{R}$.) In all test computations, ω_1 was of type B or C if $x \le 0$ and $y \ge 0$, and ω_1 was of type A or B if $x < 0$, $y \le 0$. In fact, a value of ω_1 of type C for which the latter inequalities are satisfied would contradict a conjecture which will be stated in section 5. For $x > 0$, $y > 0$ ω_1 can be of each of the types A–C, while $x \ge 0$, $y \le 0$ would contradict the assumption $\omega_1 \in R_{38}$.

Let $\omega_1 \in R_{38} \cap \mathrm{Q}[\sqrt{-11}]$ be close to α, let ρ_0, ρ_1 be such that $\rho_0/\rho_1 = \omega_1$, and let $\Sigma = \langle \rho_0, \rho_1, \rho_2, \ldots \rangle$, $\Sigma' = \langle \rho_0, \rho_1, \rho_2', \ldots \rangle$ be division chains with $\gamma_1 = (1+\sqrt{-11})/2$, $\gamma_1' = 0$, such that for $j \ge 2$ γ_j (γ_j') is an m-approximation of ω_j (ω_j'). Then the sequences $\Gamma = \langle \gamma_1, \gamma_2, \ldots \rangle$, $\Gamma' = \langle \gamma_1', \gamma_2', \ldots \rangle$ begin as follows:

$$\Gamma = \Big\langle \frac{1+\sqrt{-11}}{2}, \frac{-1+\sqrt{-11}}{2}, \frac{-1+\sqrt{-11}}{2}, \frac{3+\sqrt{-11}}{2}, \frac{1+\sqrt{-11}}{2}, \\ \frac{-3+\sqrt{-11}}{2}, \frac{3+\sqrt{-11}}{2}, \frac{-3+\sqrt{-11}}{2}, \ldots \Big\rangle$$

$$\Gamma' = \Big\langle 0, \frac{1-\sqrt{-11}}{2}, \frac{-3-\sqrt{-11}}{2}, \frac{3-\sqrt{-11}}{2}, \frac{-3-\sqrt{-11}}{2}, \ldots \Big\rangle.$$

Then the type of ω_1 will ultimately depend on the exit from this periodicity. But a full characterization of all possible exits has not been accomplished, and it may still be a non-trivial task to translate such a characterization into a geometrical criterion for ω_1 being of type A, B or C respectively.

We conclude this section by showing that the number of divisions needed if remainders of minimal norm are chosen can exceed the minimal number of divisions by an arbitrarily large amount.

THEOREM 3. *For every integer k there exists an element $\rho_0 \in \mathrm{Q}[\sqrt{-11}\,]$ with the following property: if Σ is a minimal remainder-division chain beginning with ρ_0, 1, then $l(\Sigma) - \#\mathrm{divs}(\rho_0) \geq k$.*

Proof. Looking at the sequences Σ, Σ' given at the beginning of this section, which represent the first counterexample for theorem 1 for $d = -11$, we observe that the sequence

$$\Gamma = \langle \gamma_1, \gamma_2, \gamma_3 \rangle = \left\langle \frac{1+\sqrt{-11}}{2}, \frac{-1+\sqrt{-11}}{2}, \frac{-1+\sqrt{-11}}{2} \right\rangle$$

improves

$$\Gamma' = \langle \gamma_1', \ldots, \gamma_4' \rangle = \left\langle 0, \frac{1-\sqrt{-11}}{2}, \frac{-1-\sqrt{-11}}{2}, \frac{-1-\sqrt{-11}}{2} \right\rangle.$$

More generally, $\langle \alpha + (1+\sqrt{-11}\,)/2, (-1+\sqrt{-11}\,)/2, (-1+\sqrt{-11}\,)/2 \rangle$ improves $\langle \alpha, (1-\sqrt{-11}\,)/2, (-1-\sqrt{-11}\,)/2, (-1-\sqrt{-11}\,)/2 \rangle$ for every $\alpha \in O_{-11}$. In order to prove the theorem, we show that there exist elements $\rho_0 \in \mathrm{Q}[\sqrt{-11}\,]$ such that in the minimal remainder-division chain $\Sigma = \langle \rho_0, 1, \rho_2, \ldots \rangle$ a sequence of the form $\langle \alpha, (1-\sqrt{-11}\,)/2, (-1-\sqrt{-11}\,)/2, (-1-\sqrt{-11}\,)/2 \rangle$ occurs at least k times within the quotient-sequence $\langle \gamma_1, \gamma_2, \ldots \rangle$. Suppose the first four (minimal remainder-) quotients $\gamma_1, \ldots, \gamma_4$ are α, $(1-\sqrt{-11}\,)/2$, $(-1-\sqrt{-11}\,)/2$ and again $(-1-\sqrt{-11}\,)/2$. By applying the usual method once again we find that the set of all possible values of ω_5 is the region R_{52} shown in fig. 21.

Thus γ_5 may be $(-3-\sqrt{-11}\,)/2$ or $(-1-\sqrt{-11}\,)/2$, but choosing $\gamma_5 = (-1-\sqrt{-11}\,)/2$ leaves much more options for γ_6. (To 'choose $\gamma_5 = (-1-\sqrt{-11}\,)/2$' means in fact: to construct ρ_0 such that with $\gamma_1, \ldots, \gamma_4$ as above, $\gamma_5 = (-1-\sqrt{-11}\,)/2$ will be a minimal remainder-quotient.) Indeed, if $\gamma_5 = (-1-\sqrt{-11}\,)/2$, then the region for ω_6 is the region R_{53} also shown in fig. 21. Then, if we pick $\gamma_6 = -1-\sqrt{-11}$ (away from the boundaries of R_{53}), then ω_7 can take every value in R_0^{-1}, and a minimal remainder-quotient γ_7 can take every value except 0 or ± 1. Hence a new occurrence of the subsequence $\langle \alpha, (1-\sqrt{-11}\,)/2, (-1-\sqrt{-11}\,)/2, (-1-\sqrt{-11}\,)/2 \rangle$ can start, of which $\alpha = -1-\sqrt{-11}$ was already the beginning. In short, ρ_0 is chosen such that the sequence $\langle \gamma_1, \gamma_2, \ldots \rangle$ begins with k copies of $\langle -1-\sqrt{-11}, (1-\sqrt{-11}\,)/2, (-1-\sqrt{-11}\,)/2, (-1-\sqrt{-11}\,)/2, (-1-\sqrt{-11}\,)/2 \rangle$. As indicated, the existence of such an element ρ_0 follows from the fact that with $\tilde{R}_i$ being the set of all possible values of ω_i under the assumption that $\gamma_1, \ldots, \gamma_{i-1}$ are as specified, the $\tilde{R}_i$'s follow a periodic pattern, and none of them is empty. Another way to check this claim is as follows. Compute an element $\tilde{\omega}$ (in an algebraic extension field of $\mathrm{Q}[\sqrt{-11}\,]$) with continued fraction-expansion

$$\overline{\{-1-\sqrt{-11}, (1-\sqrt{-11}\,)/2, (-1-\sqrt{-11}\,)/2, (-1-\sqrt{-11}\,)/2, (-1-\sqrt{-11}\,)/2\}}.$$

This element is found by solving the equation $\tilde{\omega} = \Omega(\tilde{\omega}, \langle -1-\sqrt{-11}, (1-\sqrt{-11}\,)/2, (-1-\sqrt{-11}\,)/2, (-1-\sqrt{-11}\,)/2, (-1-\sqrt{-11}\,)/2 \rangle)$, which is quadratic after simplification. One

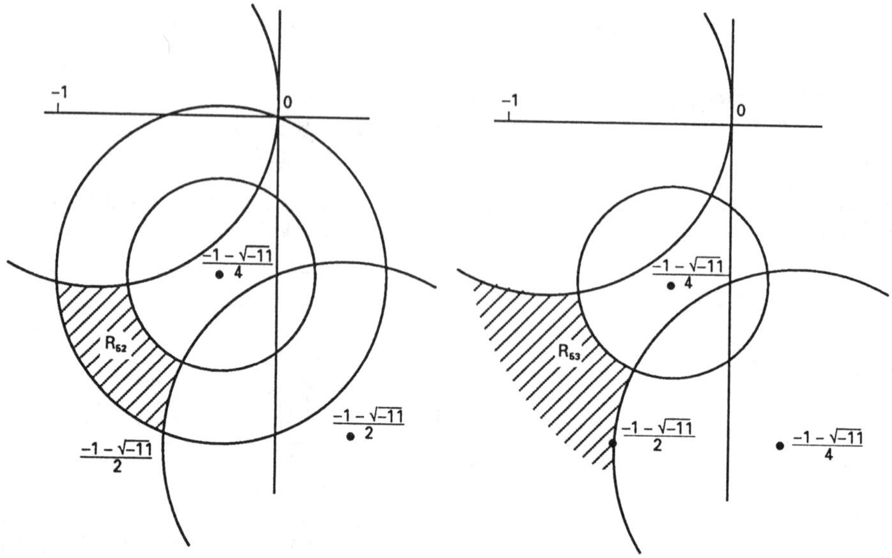

Fig. 21: regions R_{52} and R_{53}

of the two solutions, namely

$$\tilde{\omega} = -\frac{11 + \sqrt{11(2\sqrt{4645} - 43)} + \sqrt{-11}\,(19 + \sqrt{2\sqrt{4645} + 43}\,)}{44}$$

satisfies the condition that in a minimal remainder-division chain (with quotients in O_{-11}) beginning with $\tilde{\omega}, 1$, the sequence $\langle \gamma_1, \gamma_2, \ldots \rangle$, does indeed consist of infinitely many copies of $\langle -1 - \sqrt{-11}, (1 - \sqrt{-11}\,)/2, (-1 - \sqrt{-11}\,)/2, (-1 - \sqrt{-11}\,)/2, (-1 - \sqrt{-11}\,)/2 \rangle$. The desired ρ_0 is then an element of $Q[\sqrt{-11}\,]$ sufficiently close to $\tilde{\omega}$.

Let Γ be the initial segment $\langle \gamma_1, \gamma_2, \ldots \rangle$ consisting of the k copies of $\langle -1 - \sqrt{-11}, (1 - \sqrt{-11}\,)/2, (-1 - \sqrt{-11}\,)/2, (-1 - \sqrt{-11}\,)/2, (-1 - \sqrt{-11}\,)/2 \rangle$. The final step of the proof consists in constructing a sequence Γ' which improves Γ and has k terms less than Γ. The simplest way is to proceed in k steps, starting from the sequence $\Gamma' = \Gamma$; at each step one copy of $\langle -1 - \sqrt{-11}, (1 - \sqrt{-11}\,)/2, (-1 - \sqrt{-11}\,)/2, (-1 - \sqrt{-11}\,)/2, (-1 - \sqrt{-11}\,)/2 \rangle$ is replaced by $\langle (1 - \sqrt{-11}\,)/2, (-1 + \sqrt{-11}\,)/2, (-1 + \sqrt{-11}\,)/2, (3 + \sqrt{-11}\,)/2 \rangle$, beginning with the k-th copy and moving on to the first. Each time some signs must also be changed in the elements γ_i' following after the copy under consideration, as indicated in the proof of lemma 1. The reason why $(3 + \sqrt{-11}\,)/2$ appears as the last element of the sequence replacing the period of Γ' is that the d-relation implied by $\langle \gamma_1, \ldots, \gamma_4 \rangle = \langle -1 - \sqrt{-11},$

$(1-\sqrt{-11})/2, (-1-\sqrt{-11})/2, (-1-\sqrt{-11})/2\rangle$, $\langle\gamma_1', \ldots, \gamma_3'\rangle = \langle(1-\sqrt{-11})/2, (-1+\sqrt{-11})/2, (-1+\sqrt{-11})/2\rangle$ is $\omega_4' = 1-\omega_5$; this requires to replace the last element γ_5 of the period of Γ by $\gamma_4' = 1-\gamma_5$, and similarly for all other periods. At each of the k steps the length of Γ' is reduced by 1, hence at the end Γ' has the desired length. □

We note that with little extra work one could prove $l(\Sigma) - \#\text{divs}(\rho_0) = k$, not just $l(\Sigma) - \#\text{divs}(\rho_0) \geq k$, for an appropriate value of ρ_0; for this purpose one would have to show that the division chain obtained after replacing Γ by Γ' has minimal length. This is of little importance, however.

5. Conclusion

When studying the proof in sections 2–4, several questions naturally arise. Most interesting is the following: is it a coincidence that all cases work out the way they did, or is there some deeper reason, which might allow to settle all cases by a single argument? The exceptional case for $d = -11$ suggests that no such deeper reason exists. However, there may still exist an argument which covers many, though not all, of the cases involved. Consider, for instance, the case $\gamma_1 = 1$, d arbitry, and assume that $\Re\,\omega_1$ is close to $1/2$; γ_1' is 0, as always. In the borderline case, where $\Re\,\omega_1 = 1/2$, we have $\omega_1 - \gamma_1 = -\overline{\omega_1 - \gamma_1'}$, hence $\omega_2' = -\overline{\omega_2}$, which, by lemma 1 and the trivial equality $\#\text{divs}(\omega) = \#\text{divs}(\overline{\omega})$, implies $\#\text{divs}(\omega_2) = \#\text{divs}(\omega_2')$. If, on the other hand, $\Re\omega_1 < 1/2$, then the pair $(\omega_1 - 1, \overline{\omega_1 - 0})$ is obtained from a pair $(\tilde{\omega}, -\overline{\tilde{\omega}})$, $\Re\tilde{\omega} = -1/2$, by adding a real number $a < 0$ to $\tilde{\omega}$ and to $-\overline{\tilde{\omega}}$. It is conceivable that this alone implies $\#\text{divs}(1/(\omega-1)) \geq \#\text{divs}(1/\omega)$. More generally, the following may be true and provable by some direct argument, without considering many individual cases:

Conjecture: If ρ_2 and ρ_2' are remainders of ρ_0 and ρ_1, and $|\Re\,\rho_2'| \leq |\Re\,\rho_2|$, $|\Im\,\rho_2'| \leq |\Im\,\rho_2|$, then $\#\text{divs}(\rho_1/\rho_2') \leq \#\text{divs}(\rho_1/\rho_2)$.

The conjecture does indeed hold for all discriminants except possibly -11, and can be proven just like theorem 1 in these cases. Also, the theorem for $d = -3$, -4 and -8 follows from the conjecture for these discriminants. This is obvious for $d = -4, -8$, while for $d = -3$ it can be derived from the following fact: If ρ_2 and ρ_2' are remainders of ρ_0 and ρ_1, and $|\rho_2'| \leq |\rho_2|$, then there exists a unit ϵ such that $|\Re(\epsilon\rho_2')| \leq |\Re(\epsilon\rho_2)|$ and $|\Im(\epsilon\rho_2')| \leq \Im(\epsilon\rho_2)|$. But for $d = -7, -11$, the conjecture would not allow to settle the case $\gamma_1 = (1+\sqrt{d})/2$. Hence, if a short proof for the conjecture should be found, some work would still be needed to prove theorem 1 for $d = -7$. It seems very likely that the conjecture also holds for $d = -11$, but no proof has been constructed.

In the various cases from table 1, and also in the cases with unbounded look-ahead, various heuristic approaches were taken in order to find the right quotients γ_2', γ_3', ... Frequently, an example was constructed, where the current sequence of quotients γ_1, γ_2, ... applies, and γ_2', γ_3' were chosen such that Σ' became a minimal remainder-sequence. Sometimes the last of these quotients were modified, if inspection or trial and error showed that this allows to obtain the desired d-relation more quickly. Also, for $\gamma_1 = 1$, it turned out that we always had to choose $\gamma_i' = -\overline{\gamma_i}$. This is not surprising, for if $\Re\,\omega_1 = 1/2$ (the borderline case), then it is exactly this choice of γ_2', ... which shows $l(\Sigma') = l(\Sigma)$, if one follows the proof of lemma 1. A similar rule applied for the case $d = -8$, $\gamma_1 = \sqrt{-2}$. In general it is quite possible that some of the choices of γ_i' can be improved.

Is it even possible that all cases with unbounded look-ahead could actually be solved by bounded look-ahead? The answer is no, as we will show for one sample case. Consider $d = -4$, $\gamma_1 = 1$, $\gamma_2 = -2 - i$, $\gamma_j = 2(-1)^{j+1} - 2i$ for $j = 3, 4, \ldots$. As in section 3, let l be the smallest index for which $\gamma_l \neq \pm 2 - 2i$. The proof for this case could be carried out by bounded look-ahead only if the sequence $\Gamma = \langle 1, -2-i, 2-2i, \ldots, 2(-1)^{k+1} - 2i \rangle$ of length k were 0-improvable for some k. We show that this is impossible. In the investigation of the case $\gamma_1 = 1$, $\gamma_2 = -2 - i$ in section 3 it turns out that after picking $\gamma_j = \pm 2 - 2i$ for $j = 3, \ldots, m - 1$ (m arbitrary), ω_m can indeed have every possible value in R_{12} if m is odd, or every possible value in R_{13}, if m is even. (In this respect the situation differs from some other cases with unbounded look-ahead.) In particular, l can take every value ≥ 3, and every exit condition occurs for some ω_1. Consider a value of ρ_0 for which $l > k + 1$ (k as above), and for which exit condition C_{14} or C_{17} occurs. Let Σ, Σ' be the sequences constructed in the proof in section 3 for this case for this value of ρ_0, were $\gamma_1, \gamma_2, \ldots$ are as above, $\gamma_1' = 0$, $\gamma_2' = 2 - i$ etc. We have $l(\Sigma') < l(\Sigma)$. Since $\gamma_2, \gamma_3, \ldots$ are minimal remainder-quotients, it follows that $l(\Sigma') < l(\Sigma'')$ for every division chain Σ'' beginning with ρ_0, 1, $\rho_2'' = \rho_0 - 1$. This implies that $\Gamma' = \langle 0, 2 - i, -2 - 2i, \ldots, 2(-1)^k - 2i \rangle$ is not 1-improvable. Next we conclude that $\Gamma'' = \langle -1, 2 - i, -2 - 2i, \ldots, 2(-1)^k - 2i \rangle$ is not 0-improvable, for if $\langle 0, \gamma_2''', \ldots \rangle$ improves Γ'', then $\langle 1, \gamma_2''', \ldots \rangle$ would improve Γ'. Finally, the sequence Γ considered above equals $\overline{\Gamma''_{-1}}$, so the claim that Γ is not 0-improvable follows by lemma 2 (a) from section 2.

Similar proofs work for all other cases with unbounded look-ahead, except for the last one from section 3, where $\gamma_1 = \sqrt{-2}$, $\gamma_2 = 1 + \sqrt{-2}$ and where $\gamma_1 = 1 + \sqrt{-2}$, $\gamma_2 = -1$ respectively. For these two cases we do not know if bounded look-ahead might work as well.

In the case just considered, l can be large only if ω_1 is very close to $1/2 + (\sqrt{2} - 1)i/2$, more precisely: if for every m $\rho_0^{(m)}$ denotes a value of ρ_0 for which l takes the value m, then

$$\lim_{m \to \infty} \rho_0^{(m)} = \frac{1}{2} + \frac{\sqrt{2} - 1}{2} i.$$

Based on this observation, it is easy to show that in every neighborhood of $1/2 + (\sqrt{2} - 1)i/2$ there exist values of ρ_0 for which each of the following alternatives hold, where Σ, Σ' denote division chains beginning with ρ_0, 1, such that $\gamma_1 = 1$, $\gamma_1' = 0$:

(a) $l(\Sigma) = l(\Sigma') + 1$;
(b) $l(\Sigma) = l(\Sigma')$;
(c) $l(\Sigma) = l(\Sigma') - 1$.

This implies the result shown in (Rolletschek 1986), that even the slightest deviation from a minimal remainder-division chain (in a sense defined there) can cause an increase in the number of divisions. Similar results can be shown for $d = -7, -8$, based on cases with unbounded look-ahead occurring for these discriminants.

A more ambitious goal than the one pursued in this paper would be to provide for every γ_1 a complete characterization of those values of ω_1 for which each of the following alternatives hold:

(i) $\#\text{divs}(1/(\omega_1 - \gamma_1)) = \#\text{divs}(\omega_1) - 1$, that is, we can choose this value of γ_1 to obtain a shortest division chain beginning with ω_1, 1;

(ii) $\#\text{divs}(1/(\omega_1 - \gamma_1)) = \#\text{divs}(\omega_1)$;

(iii) $\#\text{divs}(1/(\omega_1 - \gamma_1)) = \#\text{divs}(\omega_1) + 1$.

Note that $\#\text{divs}(1/(\omega_1 - \gamma_1)$ cannot be $> \#\text{divs}(\omega_1)+1$, for if without loss of generality 0 is an m-quotient of ω_1 and 1, then every choice of γ_1 allows to obtain a division chain whose length exceeds a shortest division chain by 2, putting $\gamma_2 = 0$; see section 2.

Such a complete characterization is made more difficult by the cases with unbounded look-ahead. For instance, in the case $d = -4$, $\gamma_1 = 1$, the set of all ω_1 for which alternative (ii) holds is the union of an infinite set of regions, one for every value of l, where l is defined as above. We have not tried to provide a complete answer, but the proof does give a partial one; for instance, if $d = -4$, $\omega_1 \in R_0$, then choosing $\gamma_1 = 2 + i$ makes at least one extra division necessary.

In order to achieve this partial information, we have left out one possible shortcut for the proofs of theorems 1 and 2, which would make the consideration of some of the cases unnecessary. It was shown in sections 2,3 (case $\gamma_1 = 1$) that if $\rho_0 \in R_0$, then $\#\text{divs}(1/\rho_0) \leq \#\text{divs}(1/(\rho_0 - 1))$. The same conclusion can be proven with no additional effort under the assumption $\Re\,\rho_0 \leq 1/2$ instead of $\rho_0 \in R_0$, as it was done in section 4 for $d = -11$. This implies the following more general assertion, which, incidentally, is contained in the conjecture stated above:

(IV) If $\Re\rho_0 \leq \Re\gamma + 1/2$, then $\#\text{divs}(1/(\rho_0 - \gamma)) \leq \#\text{divs}(1/(\rho_0 - (\gamma+1)))$, that is, choosing $\gamma_1 = \gamma$ requires no more divisions than choosing $\gamma_1 = \gamma + 1$.

This fact allows, for instance, to reduce the case $\gamma_1 = 2 + i$ to the case $\gamma_1 = 1 + i$ for $d = -4$, and similarly for all γ_1 with $\Re\,\gamma_1 \geq 1$. Something similar can be shown for imaginary instead of real parts. But the case $\gamma_1 = (1 + \sqrt{d})/2$ still has to be dealt with for $d \equiv 1 \bmod 4$, and the proof of (IV) requires the consideration of several subcases, just as the case $\gamma_1 = 1$ did for all the discriminants. Hence this shortcut still would not change the nature of the proof essentially.

We conclude by discussing the consequences of theorems 1 and 2 for computing gcd's efficiently. Of course they play a role only for inputs of moderate size, where asymptotically fast methods do not pay. The relevance of these theorems is limited, since in general it is more important to determine the individual quotients γ_i quickly then to reduce the number of divisions by 1 or 2 in a few borderline cases. Theorem 1 does, however, have a definite consequence for gcd-computations in O_{-4} and O_{-8}: here one should definitively try to make all γ_i's minimal remainder-quotients, since they can be determined most easily (among all reasonable choices of γ_i), and also minimize the number of divisions. In an actual computation it can nevertheless happen that γ_i will not be a minimal remainder-quotient, but only due to the fact that only the leading digits of ρ_{i-1} and ρ_i are used to compute it. (See Caviness/Collins (1976) for the case $d = -4$.) The situation is not quite the same for $d = -3$ or -7. Here it might be simpler to choose the quotients γ_i differently, for instance such that

$$|\Im\,\gamma_i - \Im\,\omega_i| \leq \frac{\sqrt{|d|}}{4}, \qquad |\Re\,\gamma_i - \Re\,\omega_i| \leq \frac{1}{2}.$$

We leave it open whether the Euclidean algorithm becomes more efficient by choosing the γ_i's in this fashion, or by picking minimal remainder-quotients. The same question also applies for $d = -11$; in this case it would be a bad idea to try to make every γ_i an m-approximation for ω_i. Even if a rule for finding such m-approximations could be found, it would at least require to check several more inequalities than finding minimal remainder-quotients does.

References

Caviness, B. F., Collins, G. E. (1976). Algorithms for Gaussian integer arithmetic. In: (Jenks, R. E., ed.) *Proceedings of the 1976 Symposium on Symbolic and Algebraic Computation,* pp. 36–45. New York: Association for Computing Machinery Inc.

Cooke, George E. (1976). A weakening of the Euclidean property for integral domains and applications to algebraic number theory I. *J. reine angew. Math.* **282**, 133–156.

Hasse, H. (1964). *Vorlesungen über Zahlentheorie.* Berlin: Springer-Verlag.

Kaltofen, E., Rolletschek, H. (1985). Arithmetic in quadratic fields with unique factorization. In: (Caviness, B. F., ed.) Proceedings of the EUROCAL '85 Conference on Computer Algebra. *Springer Lec. Notes Comp. Sci.* **204**, 279–288. Revised version to appear in *Math. Comput.* **51**, October 1989.

Lazard, D. (1977). Le meilleur algorithme d'Euclide pour $K[X]$ et Z. *Comptes Rendus Acad. Sci. Paris* **284**, 1–4, with corrections in *Zbl. Math.* **341**, p. 81.

Lazard, D. (1985). Private communication.

Rolletschek, H. (1986). On the number of divisions of the Euclidean algorithm applied to Gaussian integers. *J. Symbolic Computation* **2**, 261–291.

Samuel, P. (1971). About Euclidean rings. *J. Algebra* **19**, 282–301.

Stark, H. (1967). A complete determination of complex quadratic fields of class number one. *Mich. Math. J.* **17**, 1–27.

Computing the Structure of Finite Algebras

LAJOS RÓNYAI†

Computer and Automation Institute, Hungarian Academy of Sciences, Budapest, P.O.B. 63, H-1502 Hungary

(*Received* 1 *October* 1987)

In this paper we address some algorithmic problems related to computations in finite-dimensional associative algebras over finite fields. Our starting point is the structure theory of finite-dimensional associative algebras. This theory determines, mostly in a nonconstructive way, the building blocks of these algebras. Our aim is to give polynomial time algorithms to find these building blocks, the radical and the simple direct summands of the radical-free part. The radical algorithm is based on a new, tractable characterisation of the radical. The algorithm for decomposition of semisimple algebras into simple ideals involves (and generalises) factoring polynomials over the ground field.

Next, we study the problem of finding zero divisors in finite algebras. We show that this problem is in the same complexity class as the problem of factoring polynomials over finite fields. Applications include a polynomial time Las Vegas method to find a common invariant subspace of a set of linear transformations as well as an explicit isomorphism between a given finite simple algebra and a full matrix algebra over a finite field.

1. Introduction

Our main objects of study are associative algebras. A is an *associative algebra* over the field F if:

(a) A is a linear space over F;
(b) A is equipped with a binary F-bilinear operation $*$ (i.e. the multiplication);
(c) the multiplication is associative:

$$x * (y * z) = (x * y) * z \text{ holds for every } x, y, z \in A.$$

As is usual, we write xy instead of $x * y$. In this paper we shall consider finite-dimensional algebras only, i.e. we assume that $dim_F A$ is finite. We shall use the terms and notions *commutative algebra, subalgebra, (left) ideal, factor algebra, homomorphism, A-module* in the standard way, cf. Herstein (1968) and Pierce (1982).

An algebra A is *simple* if A has only trivial ideals (i.e. (0) and A) and $AA \neq (0)$. We say that A is the direct sum of its (left) ideals $A_1, \ldots, A_k$ (written as $A_1 \oplus \cdots \oplus A_k$) if A is the direct sum of these linear subspaces.

EXAMPLES.
(a) If the field K is a finite algebraic extension of the field F then K is a finite-dimensional simple and commutative algebra over F.
(b) $M_n(F)$, the algebra of all n by n matrices over the field F. $M_n(F)$ is a simple algebra over F and $dim_F M_n(F) = n^2$.

† Research partially supported by Hungarian National Foundation for Scientific Research, Grant 1812.

(c) Subalgebras of $M_n(F)$, i.e. linear subspaces of $M_n(F)$ closed under matrix multiplication.

The latter examples tend to be typical as the following well-known representation theorem shows.

REPRESENTATION THEOREM. *Let A be an algebra over the field F and suppose that* $\dim_F A = n$. *Then A is isomorphic to a subalgebra of* $M_{n+1}(F)$. *Moreover, if A has an identity element, then A is isomorphic to a subalgebra of* $M_n(F)$.

If A has no identity element then we can adjoin one using the Dorroh extension (cf. Kertész, 1987, p. 43) with the ground field. This process increases the dimension of A by one.

The Representation Theorem is easily proved using the *regular representation* of A. For each $x \in A$ we define the linear map $R_x: A \to A$ as $R_x(y) = xy$ for every $y \in A$. It is easy to see that R is an algebra homomorphism of A to the algebra of linear transformations of the linear space A. If A has an identity element then R is injective.

An element $x \in A$ is *nilpotent* if $x^m = 0$ for some positive integer exponent m. An element $x \in A$ is *strongly nilpotent* if xy is nilpotent for every $y \in A$. The *radical* of A, denoted by $Rad(A)$, is defined as

$$Rad(A) = \{x \in A;\ x \text{ is strongly nilpotent}\}.$$

It is not difficult to see that $Rad(A)$ is an ideal of A and the factor algebra $A/Rad(A)$ has no strongly nilpotent elements except 0. Algebras having no strongly nilpotent elements except 0 are called *semisimple algebras.* Semisimple algebras admit a very nice structure theorem due to Wedderburn (cf. Herstein, 1968; Pierce, 1982).

WEDDERBURN STRUCTURE THEOREM. *Suppose that A is a finite-dimensional semisimple algebra over the field F. Then A can be expressed as a direct sum of simple algebras*

$$A = A_1 \oplus A_2 \oplus \cdots \oplus A_k, \tag{1.1}$$

where the A_i *are the only minimal nontrivial ideals of A. Moreover, each* A_i *is isomorphic to a full matrix algebra* $M_{n_i}(F_i)$ *where* F_i *is a not necessarily commutative extension field of F.*

We remark that the theorem of Wedderburn on finite division rings (cf. Herstein, 1968, p. 71) implies that if F is finite then the fields F_i are commutative.

We need some facts on Lie algebras. The results quoted can be found in Jacobson (1962) and Humphreys (1980). A *Lie algebra* over the field F is a linear space L over F equipped with an F-bilinear binary operation [ ] such that the following identities are valid for any $x, y, z \in L$.

(a) $[xx] = 0$,
(b) $[[xy]z] + [[yz]x] + [[zx]y] = 0$.

As in the associative case, one can speak about *subalgebras, ideals, factor algebras* and *homomorphisms* of Lie algebras. The *derived series* of L is the sequence $L^{(j)}$ of ideals of L where $L^{(0)} = L$ and $L^{(i+1)} = [L^{(i)}L^{(i)}]$ for $i > 0$. L is *solvable* if $L^{(n)} = (0)$ for some natural number n. If L is finite dimensional over F then it has a unique maximal solvable ideal $R(L)$, the *radical* of L.

The *descending central series* of L is the sequence L^j of ideals of L where $L^0 = L$ and $L^{i+1} = [LL^i]$ for $i > 0$. L is *nilpotent* if $L^n = (0)$ for some natural number n. If L is finite dimensional over F then it has a unique maximal nilpotent ideal $N(L)$, the *nilradical* of L.

EXAMPLE. For $A, B \in M_n(F)$ let $[AB] = AB - BA$ (i.e. the additive commutator). It is easy to see that this operation is F-bilinear and satisfies requirements (a) and (b), so if a subspace L is closed under the operation [ ] then L can be considered as a Lie algebra. Lie algebras obtained in this way are called *linear Lie algebras.*

There is a straightforward but usually not faithful representation of abstract Lie algebras as linear Lie algebras. For an $x \in L$ let $ad(x): L \to L$ be the linear map for which $ad(x)y = [xy]$. The map $x \mapsto ad(x)$ is a Lie algebra homomorphism of L to the linear Lie algebra of all linear transformations of L. We remark that a deep theorem of Iwasawa and Ado (Jacobson, 1962, chap. 6) states that every finite-dimensional Lie algebra is actually isomorphic to a linear Lie algebra.

When speaking of algorithms, one has to specify the input of the algorithmic questions considered. An algebra (associative or Lie) can be given by a collection of *structure constants.* If A is an algebra over a field F and $a_1, a_2, \ldots, a_n$ is a linear basis of A then multiplication $\circ$ can be specified by representing the products $a_i \circ a_j$ as linear combinations of the a_i

$$a_i \circ a_j = \gamma_{ij1} a_1 + \cdots + \gamma_{ijn} a_n.$$

The coefficients $\gamma_{ijk} \in F$ are called structure constants. In this paper algebras are considered to be given as a collection of structure constants. As a special case, F is an algebra over its prime field, therefore F can also be inputted with structure constants. Finite fields can be (and often are) specified by giving the minimal polynomial f of a single generating element α over the prime field $GF(p)$. Notice that this representation is a special case of the representation with structure constants. The coefficients of f give the structure constants with respect to the basis $1, \alpha, \alpha^2, \ldots, \alpha^{n-1}$ of F where $n = dim_{GF(p)} F$.

We remark that algebras may be given as matrix algebras. In this case it suffices to specify a set of matrices which generates the algebra. Our results are applicable in this setting as well, since from this representation one can efficiently find a basis of the algebra and then the structure constants with respect to this basis. In the case of associative algebras the regular representation gives an efficient method to obtain a matrix representation from structure constants. We mention here that the methods do not seem to be directly applicable when the algebras are given by generators and relations.

Our aim is to give algorithms which run in time polynomial in terms of the input size (cf. Hopcroft & Ullman, 1979, Chapters 12–13). Modulo p residue classes have size $\lceil \log_2(p+1) \rceil$. The size of composite objects (matrices, vectors, etc.) can be obtained by adding up the sizes of their parts. Polynomial time algorithms are known (Knuth, 1981; Lidl & Niederreiter, 1983) for the basic seminumerical problems (such as arithmetical operations in finite fields, polynomial arithmetic over finite fields). Also, we have efficient methods to solve systems of linear equations over finite fields. A central algorithmic problem in this context is the problem of factoring polynomials over finite fields. A deterministic method was given by Berlekamp (1968; see also Lidl & Niederreiter, 1983). The time complexity of this algorithm is a polynomial in the parameters p, s and $deg(f)$ where $f \in GF(q)[x]$ is the polynomial to be factored and $q = p^s$, p prime. Note that the input size in this case is $O((1 + deg(f)) \log q)$, consequently the running time of the above method is not polynomial in the input size. The problem can be solved in polynomial time if we allow randomisation. The first such method was given by Berlekamp (1970) [see also Ben Or (1981); Camion (1983); Cantor & Zassenhaus (1981); Rabin (1980)]. Actually, this method belongs to a special kind of randomised algorithms. For an arbitrary input it either gives a correct solution or, with a small probability, reports failure (i.e. never gives an incorrect answer). Such methods are called *Las Vegas algorithms.* The term was introduced by Babai (1979). We will use both deterministic and Las Vegas factoring algorithms. To handle this situation, the following definition will be convenient. By an *f-algorithm* we mean an algorithm which uses an oracle (subroutine) to factor polynomials over finite fields. The cost of a call of this oracle is the length of the input of the call.

In section 2 we give a deterministic polynomial time algorithm to compute the radical of finite algebras. This method will depend on a new, algorithmic characterisation of the radical. This result is then applied to the computation of the (solvable) radical and the nilradical of finite Lie algebras.

In section 3 a polynomial time f-algorithm is described to find the Wedderburn decomposition (1.1) of a finite semisimple algebra A. The algorithm is based on the work of Friedl (1983), cf. Friedl & Rónyai (1985).

In section 4 a polynomial time f-algorithm is given to find zero divisors in a finite algebra A, i.e. nonzero elements $x, y \in A$ such that $xy = 0$. It may come as a surprise that the case $A \cong M_n(F)$ presents most of the difficulties here. The main algorithm of section 3 can be viewed as a special case of the zero divisor algorithm.

In section 5 we give some applications of the zero divisor algorithm. First, we develop an important auxiliary procedure to find explicit isomorphisms of full matrix algebras. This enables us to find a common invariant subspace for a set of linear operators and to express a finite module over a finite semisimple algebra as a direct sum of simple submodules. These algorithms also run in polynomial time, as f-algorithms. Finally, we derive a deterministic polynomial time procedure to solve the following computational problem on permutation groups, raised by W. M. Kantor.

Let $G \leq S_n$ and let $K < H$ be normal subgroups of G such that H/K is an elementary Abelian p-group for some prime p. The groups G, H and K are given by generating sets. Find a normal subgroup L of G, minimal subject to the conditions $K < L \leq H$.

Our primary objective is to exhibit (deterministic or randomised) polynomial time algorithms to solve the problems. Beyond that we do not address the efficiency of the methods. In fact, in many cases these results can only be regarded as first steps towards obtaining good algorithms.

2. Computing the Radical

In this section we give a polynomial time algorithm to compute the radical of finite associative algebras. The method can be applied to the problem of computing the nilradical and the solvable radical of finite Lie algebras. The main algorithmic problem we consider here is the following.

Given a finite-dimensional associative algebra A over the field $GF(q)$ by a collection of structure constants, find a basis of $Rad(A)$, the radical of A in time polynomial in the input size, i.e. the parameters $dim_{GF(q)}A$ and $\log q$.

We claim that it suffices to solve the problem for algebras over a prime field $GF(p)$. Indeed, A can be considered as an algebra over $GF(p)$ and the radical, as the set of strongly nilpotent elements of A, does not depend on the ground field considered. Also, this change does not affect the input size. Finally, from a $GF(p)$ basis of $Rad(A)$ we can easily obtain a basis over $GF(q)$.

We remark that the problem of finding the radical over fields of characteristic zero is equivalent to solving a system of linear equations over the ground field. One can use the following theorem of Dickson (1923, pp. 106–108) on the radical of matrix algebras.

DICKSON'S THEOREM. *Let A be a finite-dimensional algebra of matrices over a field F such that $charF = 0$. Then*

$$Rad(A) = \{x \in A;\ Tr(xy) = 0 \text{ for every } y \in A\}.$$

This result shows that if $charF = 0$ then $Rad(A)$ can be obtained by solving a system of linear equations over F.

EXAMPLE. Let A denote the algebra of all p by p scalar matrices over the finite prime field $GF(p)$. Clearly, A is semisimple but the trace form identically vanishes on A. This shows that the condition on the characteristic of the field in Dickson's theorem cannot be dropped.

We return to the finite case. Let p be a prime and let $F = GF(p)$ and suppose that A is a subalgebra of $M_n(F)$ where $dim_F A = n$ or $dim_F A = n-1$. Using the regular representation of A, we can efficiently achieve this situation. We define the natural number l by the following inequalities: $p^l \leq n < p^{l+1}$. Let B denote the set of matrices $A \cup \{1_n\}$ where 1_n is the identity element of $M_n(F)$.

Our main objective is to define a sequence of ideals $I_{-1}, I_0, \ldots, I_l$ of A and a sequence of functions $g_i: I_{i-1} \rightarrow F$, $i = 0, 1, \ldots, l$ with the following properties.

(1) $I_{-1} = A$ and $I_l = Rad(A)$.
(2) g_i is an F-linear function for $i = 0, \ldots, l$.
(3) $I_i = \{x \in I_{i-1};\ g_i(xy) = 0$ for every $y \in B\}$.
(4) $g_i(x)$ can be computed for any $x \in A$ in time polynomial in n and $\log p$.

These properties immediately imply that from a basis of I_{i-1} we can obtain a basis of I_i by solving a system of linear equations over F. By linearity of the functions g_i and by property (4) the coefficients of this sytem can be obtained in polynomial time. If we start with $I_{-1} = A$ then we can find $Rad(A)$ using $l+1 = O(\log n)$ iterations, therefore (a basis of) $Rad(A)$ over F can be obtained using $(n + \log p)^{O(1)}$ bit operations.

Before defining these ideals and linear functions we need some preparation.

Let M_n denote the ring of n by n matrices over the integers and ϕ denote the ring homomorphism from M_n to $M_n(F)$ induced by the $Z \rightarrow F$ epimorphism (i.e. ϕ denotes the *mod p* reduction of integral matrices). Matrices over Z will be denoted by capitals (C, D, X, Y), the corresponding lower-case letters denote matrices over F.

We want to speak about $Tr(c^{p^i})(mod\ p^{i+1})$ where $c \in M_n(F)$ by simply choosing an arbitrary integral matrix $C \in M_n$ for which $\phi(C) = c$ and taking $Tr(C^{p^i})(mod\ p^{i+1})$. This procedure is justified by the following lemma.

LEMMA 2.1. *Let $C, D \in M_n$ and suppose that $\phi(C) = \phi(D)$. Then for any natural number i we have*

$$Tr(C^{p^i}) \equiv Tr(D^{p^i})(mod\ p^{i+1}).$$

PROOF. The statement is trivial for $i = 0$, therefore we may assume that $i > 0$. Let

$P = D - C$. It is clear that every entry of P is divisible by p. First we notice that if $B_1, \ldots, B_k$ are integral matrices and m of them are equal to P, then every entry of the product matrix $B = B_1 \cdots B_k$ is divisible by p^m, therefore $Tr(B)$ is also divisible by p^m.

We expand the right-hand side of the congruence stated. We obtain

$$Tr(D^{p^i}) = Tr((C+P)^{p^i}) = \sum Tr(Z_1 Z_2 \cdots Z_{p^i}),$$

where Z_j is either C or P and the summation ranges over all such products. Now let $G = \langle \pi \rangle$ denote the cyclic group of order p^i. We define an action of G on the words $Z_1 \cdots Z_{p^i}$ by setting

$$\pi(Z_1 Z_2 \cdots Z_{p^i}) = Z_{p^i} Z_1 Z_2 \cdots Z_{p^i - 1},$$

i.e. π acts as a cyclic shift. Clearly, if V and W are two words from the same G-orbit then for the corresponding products (denoted by the same letters) we have $Tr(V) = Tr(W)$ because of the identity $Tr(XY) = Tr(YX)$, $X, Y \in M_n$. If the orbit of the product V contains p^j elements then the contribution of this orbit to the sum is $p^j Tr(V)$. In this case π^{p^j} leaves V fixed, therefore V can be obtained as the p^{i-j}th power of its first p^j factors.

If V as a word is not C^{p^i} then at least p^{i-j} of the matrices appearing in the word V must be equal to P. Now using the trivial inequality $p^{i-j} \geq i-j+1$ we obtain that every entry and hence the trace of V is divisible by p^{i-j+1}, therefore the contribution of the orbit is divisible by p^{i+1}. Finally, we observe that the word C^{p^i} forms a one element orbit, proving the lemma. □

The next lemma provides a tool for inductive arguments.

LEMMA 2.2. *Let H be a multiplicatively closed subset of M_n and let k be a positive integer. Suppose that $Tr(X^{p^i})$ is divisible by p^{i+1} for every $X \in H$ and $0 \leq i < k$. Then we have*

$$Tr((X+Y)^{p^k}) \equiv Tr(X^{p^k}) + Tr(Y^{p^k}) \pmod{p^{k+1}}$$

for every $X, Y \in H$.

PROOF. This time we expand the left-hand side of the congruence. We obtain

$$Tr((X+Y)^{p^k}) = \sum Tr(Z_1 Z_2 \cdots Z_{p^k}),$$

where Z_j is either X or Y and the summation ranges over all of the 2^{p^k} such products. Let $G = \langle \pi \rangle$ denote the cyclic group of order p^k. Again, the elements of G act on the words $Z_1 Z_2 \cdots Z_{p^k}$ as cyclic shifts. If the orbit of the product V contains p^j elements then the contribution of this orbit to the sum is $p^j Tr(V)$. In this case, as in Lemma 2.1, the matrix V can be obtained as the p^{k-j}th power of its first p^j factors. This prefix is denoted by U. Clearly, we have $U \in H$ and if $j \neq 0$ then our assumptions imply $Tr(V) \equiv 0 \pmod{p^{k-j+1}}$. The sum of the orbit of V is divisible by p^{k+1}. The one element orbits of G correspond to the right-hand side of the congruence stated. □

Let $f \in F[x]$ be a monic polynomial:

$$f(x) = x^n + a_1 x^{n-1} + \cdots + a_n. \qquad (*)$$

Let $\alpha_1, \ldots, \alpha_n$ be the roots of f (in a suitable extension of F) and put

$$s_i = \alpha_1^i + \alpha_2^i + \cdots + \alpha_n^i, \quad i = 1, 2, \ldots, n.$$

The elements s_i can be expressed in terms of the coefficients of f, using Newton's identities

$$s_1 + a_1 = 0$$
$$s_2 + a_1 s_1 + 2a_2 = 0$$
$$\vdots$$
$$s_n + a_1 s_{n-1} + \cdots + a_{n-1} s_1 + n a_n = 0.$$

Using these formulae we can establish a trace condition for nilpotence. We recall that l is the unique integer determined by the inequalities $p^l \le n < p^{l+1}$.

LEMMA 2.3. *Let H be a multiplicatively closed subset of M_n and suppose that $Tr(X^{p^l}) \equiv 0 (mod\ p^{l+1})$ for every $X \in H$. Then $\phi(X)$ is nilpotent for every $X \in H$.*

PROOF. It suffices to show that $\phi(X)^{p^l} = \phi(X^{p^l})$ is nilpotent. Let

$$f(x) = x^n + a_1 x^{n-1} + \cdots + a_n$$

denote the characteristic polynomial (with leading coefficient 1) of the matrix $Y = X^{p^l}$ over the rationals. Clearly, $\phi(Y)$ is nilpotent if and only if $a_i \equiv 0 (mod\ p)$ for $1 \le i \le n$. We shall use Newton's identities for the polynomial f. Using the facts that $s_i = Tr(Y^i) = Tr((X^i)^{p^l})$ and that $X^i \in H$, we obtain that s_i is divisible by p^{l+1}. Now from Newton's identities we see that $ia_i \equiv 0 (mod\ p^{l+1})$ for $1 \le i \le n$. The definition of l implies that i is not divisible by p^{l+1}, hence a_i is divisible by p for $1 \le i \le n$ and the statement follows. □

Next, we prove a counterpart of Lemma 2.3, a necessary condition for nilpotence.

LEMMA 2.4. *Let $X \in M_n$ be a matrix such that $\phi(X)$ is nilpotent. Then*

$$Tr(X^{p^i}) \equiv 0 (mod\ p^{i+1})$$

holds for every natural number i.

PROOF. As $\phi(X)$ is nilpotent, it is similar over F to a strictly upper triangular matrix. Or, expressing this in terms of integral matrices, there exist $C, D, P, R, U \in M_n$ such that

$$CXD = U + P, \quad DC = I + R, \quad U^n = 0 \quad \text{and} \quad \phi(P) = \phi(R) = 0,$$

where I is the identity matrix (in M_n) and 0 denotes the zero matrix of M_n and $M_n(F)$ as well.

Using Lemma 2.1 we have

$$0 \equiv Tr(U^{p^i}) \equiv Tr((U+P)^{p^i}) = Tr((CXD)^{p^i}),$$

where the congruence is *mod* p^{i+1}. We have also

$$Tr((CXD)^{p^i}) = Tr((DCX)^{p^i}) = Tr((X+RX)^{p^i}).$$

Observing that $\phi(RX) = 0$, we can use Lemma 2.1 again

$$Tr((X+RX)^{p^i}) \equiv Tr(X^{p^i})(mod\ p^{i+1}).$$

By combining these equalities and congruences we obtain that $Tr(X^{p^i})$ is divisible p^{i+1}. □

After these preparations we can define the ideals I_j of the matrix algebra $A \le M_n(F)$. Let

$I_{-1} = A$ and for $0 \le i \le l$ let

$$I_i = \{x \in A;\ Tr((xy)^{p^j}) \equiv 0 (mod\ p^{j+1}) \text{ for every } y \in B \text{ and } 0 \le j \le i\}.$$

Here, $B = A \cup \{I\}$, where I is the identity matrix of $M_n(F)$ and $Tr(a^{p^j})(mod\ p^{j+1})$ is defined for $a \in M_n(F)$ by choosing an integral matrix $A \in M_n$ for which $\phi(A) = a$ and taking $Tr(A^{p^i})(mod\ p^{i+1})$. This procedure is justified by Lemma 2.1.

From the definition it is immediate that

$$A = I_{-1} \supseteq I_0 \supseteq \cdots \supseteq I_l.$$

Now we can show that these subsets are actually ideals of A.

THEOREM 2.5. *I_m is an ideal for A for $m = -1, 0, 1, \ldots, l$. Moreover, we have $I_l = Rad(A)$.*

PROOF. The first statement is obvious for $m = -1$, so we may suppose that $m \ge 0$. The definition of I_m immediately implies that if $x \in I_m$ and $u \in A$ then $xu \in I_m$. The relation $ux \in I_m$ is obtained from the following identity

$$Tr(((UX)Y)^k) = Tr((X(YU))^k), \quad U, X, Y \in M_n,$$

where k is a nonnegative integer. We have established that I_m is a semigroup ideal of A. Now we show that I_m is an additive subgroup of A. This is clear for $m = 0$ because Tr is an additive function. Now let $m > 0$. As I_m is multiplicatively closed, this is true for its preimage $J_m = \phi^{-1}(I_m)$. We shall apply Lemma 2.2 with $H = J_m$. Let $X, Y \in J_m$ and let $U \in M_n$ such that $\phi(U) \in B$. Now for $0 \le k \le m$ we have

$$\begin{aligned} Tr(((X+Y)U)^{p^k}) &= Tr((XU+YU)^{p^k}) \\ &\equiv Tr((XU)^{p^k}) + Tr((YU)^{p^k}) \equiv 0 \ (mod\ p^{k+1}), \end{aligned}$$

where the first congruence follows from the additivity of Tr for $k = 0$ and from Lemma 2.2 for $k > 0$. As for the last congruence, $\phi(X), \phi(Y) \in I_m$ imply that

$$Tr((XU)^{p^k}) \equiv Tr((YU)^{p^k}) \equiv 0 \ (mod\ p^{k+1}).$$

We conclude that I_m is an additive subgroup, therefore an ideal of A.

Finally, we show that $I_l = Rad(A)$. Indeed, if $x \in Rad(A)$ then xy is nilpotent for every $y \in B$. Let U be an integral matrix such that $\phi(U) = xy$. Then by Lemma 2.4 we have

$$Tr((xy)^{p^i}) \equiv Tr(U^{p^i}) \equiv 0 \ (mod\ p^{i+1}),$$

where i is an arbitrary natural number, showing that $x \in I_l$. It remains to see that $I_l \le Rad(A)$. This immediately follows from Lemma 2.3 if we put $H = \phi^{-1}(I_l)$. The proof is complete. □

For $0 \le i \le l$ we define the functions $f_i : M_n \to Q$ as follows

$$f_i(X) = \frac{Tr(X^{p^i})}{p^i}.$$

If $X \in \phi^{-1}(I_{i-1})$ then $f_i(X)$ is an integer and if $X, Y \in \phi^{-1}(I_{i-1})$ then we have

$$f_i(X+Y) \equiv f_i(X) + f_i(Y) \ (mod\ p). \tag{2.1}$$

The congruence (2.1) is immediate for $i = 0$ and follows from Lemma 2.2 for $i > 0$. Next, we define the functions $g_i : I_{i-1} \to F$, $i = 0, \ldots, l$ as

$$g_i(x) = f_i(X) \ (mod\ p),$$

where X is an arbitrary integral matrix for which $\phi(X) = x$. To justify this definition, let $X, Y \in M_n$ be such that $\phi(X) = \phi(Y) = x$. Then by Lemma 2.1

$$Tr(X^{p^i}) \equiv Tr(Y^{p^i}) \ (mod\ p^{i+1}),$$

$X, Y \in \phi^{-1}(I_{i-1})$ implies that p^i divides both sides of the congruence, therefore

$$f_i(X) \equiv f_i(Y) \ (mod\ p)$$

follows.

The main properties of the functions g_i are expressed in the following.

THEOREM 2.6. *For $i = 0, 1, \ldots, l$ we have*

(*i*) $g_i: I_{i-1} \to F$ *is an F-linear function;*
(*ii*) $I_i = \{x \in I_{i-1};\ g_i(xy) = 0$ *for every* $y \in B\}$;
(*iii*) *For a given* $x \in I_{i-1}$, $g_i(x)$ *can be computed using* $(\log p + n)^{O(1)}$ *bit operations.*

PROOF.

(i) Is an immediate consequence of (2.1);
(ii) this statement is a simple reformulation of the definition of I_i. Indeed, $g_i(xy) = 0$ if and only if $Tr((xy)^{p^i})$ is divisible by p^{i+1};
(iii) in view of the definition of g_i, it suffices to compute the trace of a power of an integral matrix modulo p^{l+1}. Using the fact that the exponent in question is at most n and that the entries of the matrix are from the interval $[0, p]$, the statement follows. □

With Theorems 2.5 and 2.6 we completely proved the properties (1)–(4) of the ideals I_i and functions g_i described at the beginning of this section. We have the following.

THEOREM 2.7. *Let A be an n-dimensional algebra over the field GF(q) given by a collection of structure constants. Then (a basis of) Rad(A) can be computed in time polynomial in n and* $\log q$. □

To conclude this section, we outline how finding the nilpotent and the solvable radical of a Lie algebra can be reduced to the associative case. First, we deal with the nilradical. We shall use a theorem of Jacobson (1962). Let L be a finite-dimensional Lie algebra over the field K.

JACOBSON'S THEOREM. *Let A denote the associative (matrix-) algebra generated by ad(L), the adjoint representation of L. Then an element* $x \in L$ *is in the nilradical N(L) if and only if* $ad(x) \in Rad(A)$.

This result settles the problem of computing $N(L)$ if K is a finite field or an algebraic number field. Indeed, we can compute a basis of A, and using the results of this section we can compute $Rad(A)$ as well. By solving a system of linear equations we can compute the intersection of the subspaces $ad(L)$ and $Rad(A)$.

Corollary 2.8. *Let L be a finite dimensional Lie algebra over the field K, where K is either finite or an algebraic number field, given by a collection of structure constants. Then the nilradical N(L) can be computed in time polynomial in the input size.* □

Now we turn to the problem of computing the solvable radical $R(L)$. Over fields of characteristic zero, Beck, Kolman & Steward (1977) have given an efficient algorithm to compute $R(L)$. Their method is based on a characterisation of $R(L)$ using the Killing form, which is similar to Dickson's theorem. This characterisation (as well as Dickson's theorem) breaks down over fields of positive characteristic.

If K is finite then the problem of computing $R(L)$ can be, at least in a theoretical sense, efficiently reduced to the problem of computing $N(L)$. Indeed, we have $N(L) \leq R(L)$ and if $N(L) = (0)$ then $R(L) = (0)$, because the next to last element of the derived series of $R(L)$ is an abelian, therefore nilpotent ideal of L.

One can define the sequence L_i as follows: let $L_0 = L$ and if $N(L_i) \neq (0)$ then let $L_{i+1} = L_i/N(L_i)$, otherwise L_{i+1} is not defined. This sequence of Lie algebras has at most $dim_K L + 1$ elements. By Corollary 2.8, these algebras can be computed in polynomial time if K is finite. If L_j is the last element of the sequence then $L_j \cong L/R(L)$ and we can easily produce a basis for $R(L)$ by keeping track of the preimages of the ideals we factored out during the process.

Corollary 2.9. *Let L be an n dimensional Lie algebra over GF(q), given by structure constants. Then (a basis of) the solvable radical R(L) can be computed in time polynomial in n and* log *q.* □

3. Finding the Simple Components of Semisimple Algebras

Our aim here is to find the minimal ideals, i.e. the Wedderburn decomposition of a finite semisimple associative algebra. We describe a polynomial time f-algorithm to solve this problem. This method was given by Friedl & Rónyai (1985), based on Friedl (1983). The algorithm is presented here because it is a major building block of our subsequent methods. A minor improvement of the original version is also included.

The input of the problem is a finite-dimensional semisimple associative algebra A over the field $GF(q)$ ($q = p^s$, p prime), inputted as a collection of structure constants. By the Wedderburn structure theorem

$$A = A_1 \oplus A_2 \oplus \cdots \oplus A_k,$$

where $A_1, A_2, \ldots, A_k$ are the minimal ideals of A. We give an f-algorithm running in time polynomial in $dim_{GF(q)} A$ and $\log q$ to find bases for the ideals A_i.

The problem can be reduced in polynomial time to the case when A is commutative. One can consider the centre of A

$$Z(A) \overset{\text{def}}{=} \{x \in A;\ xy = yx \text{ for every } y \in A\}.$$

A basis of the algebra $Z(A)$ can be obtained by solving a system of linear equations over $GF(q)$, since $x \in Z(A)$ if and only if $xa_i = a_i x$, $i = 1, \ldots, n$ holds where $a_1, \ldots, a_n$ is a basis of A over $GF(q)$. $Z(A)$ is a semisimple associative algebra and its Wedderburn decomposition relates nicely to that of A:

$$Z(A) = Z(A_1) \oplus Z(A_2) \oplus \cdots \oplus Z(A_k).$$

Knowing the subalgebras $Z(A_i)$, we can easily compute the ideals A_i because of the relation $A_i = Z(A_i)A$ (which, in fact, shows that A_i is the ideal of A generated by $Z(A_i)$). A basis of the product $Z(A_i)A$ can be obtained by simply selecting a maximal linearly independent set from the set of all possible products $b_j a_r$ where b_j and a_r run through a basis if $Z(A_i)$ and A, respectively.

From now on we shall assume that A is a commutative semisimple algebra over $GF(q)$. The method is an iteration which goes sequentially through a basis $a_1, \ldots, a_n$ of A. It either finds a decomposition of A into a direct sum of two smaller ideals $A = I \oplus J$ or concludes that A is a field, therefore direct irreducible. In the former case this cutting procedure can be applied to the ideals I and J and so on, since they are also semisimple commutative algebras and their ideals are also ideals of A. Thus, a call of the cutting procedure either concludes that an ideal is minimal or obtains a finer direct decomposition of A.

The cutting procedure works as follows. We consider a basis $a_1, \ldots, a_n$ of A. After processing the elements $a_1, \ldots, a_i$, the loop invariant is that F_i, the subalgebra generated by $a_1, \ldots, a_i$ is a field ($F_0 = GF(q)$). If $i = n$ then we have proved that A is a field, therefore it has no proper ideals. If $i < n$ then we compute the minimal polynomial f of the element $b = a_{i+1}$ over the field F_i. This involves finding the first linear dependence over F_i of the elements $1_A, b, b^2, \ldots, b^n$. Next, we factor f into irreducible factors over F_i (this is where we have to call the factoring oracle). If f is irreducible, then $F_i(a_{i+1})$ is a field. In this case we put $F_{i+1} = F_i(a_{i+1})$ and the ith step is finished. If f is reducible then f can be written as $f = gh$, where g and h are nonconstant relatively prime polynomials because A is a direct sum of fields. Now we put $I = Ag(a_{i+1})$ and $J = Ah(a_{i+1})$. I and J are obviously proper ideals of A, and using the fact that g and h are relatively prime, one can easily see that $A = I \oplus J$.

We have the following result.

THEOREM 3.1 (Friedl & Rónyai, 1985). *Let A be a finite dimensional semisimple associative algebra over the field $GF(q)$ ($q = p^s$, p prime), given by a collection of structure constants. Then there exists an f-algorithm running in time polynomial in $\dim_{GF(q)} A$ and $\log q$ to find bases for the minimal ideals of A (i.e. to find the Wedderburn decomposition of A).*

By putting the polynomial time Las Vegas method (Berlekamp, 1970) and the deterministic method (Berlekamp, 1968) into the oracle for factoring polynomials over finite fields, we obtain the following.

COROLLARY 3.2. *The minimal ideals of the algebra A above can be found by a Las Vegas algorithm running in time polynomial in $\dim_{GF(q)} A$ and $\log q$. Similarly, the minimal ideals of A can be found by a deterministic method running in time polynomial in $\dim_{GF(q)} A$ and q.*

REMARKS.

1. With some modifications, the algorithm of Theorem 3.1 works over algebraic number fields as well. One has to control the sizes of the intermediate results obtained from the repeated calls of the cutting procedure (Friedl & Rónyai, 1985).

2. From a practical point of view, a modified version of the above method seems to be more promising. A is a semisimple commutative algebra, hence the minimal ideals A_i are fields. These ideals contain a copy of the prime field $GF(p)$. We infer that A contains, as a subalgebra, a direct sum of k copies of $GF(p)$. This subalgebra B is the set of the fixed points of the map $x \mapsto x^p$. This is a $GF(p)$ linear map, consequently B can be computed by

solving a system of linear equations over $GF(p)$. Again, it suffices to compute the decomposition of B, for the minimal ideals of B generate the minimal ideals of A. Now during the cutting procedure we always work over the prime field (i.e. F_i never grows) and the minimal polynomials split into linear factors in the prime field. Note that this reduction is essentially the same as Berlekamp's reduction (1968) of factoring polynomials to finding roots in the prime field.

4. Simple Algebras over Finite Fields

The main task of the cutting procedure in section 3 was to find *zero divisors* in the algebra A, i.e. nonzero elements $x, y \in A$ such that $xy = 0$. There we solved this problem in the special case when A was a semisimple but not simple algebra. Here we consider this algorithmic problem in a broader context.

We have a finite associative algebra A given as a collection of structure constants, find a pair of zero divisors in A if there are any.

We give a polynomial time f-algorithm to solve the problem. As it turns out, the most important special case is when A is simple. Our solution of this case is based on an almost constructive proof of Wedderburn's theorem on finite division algebras presented in Herstein (1968). We adapt the proof to full matrix algebras and replace the purely existential steps by constructive ones. Let $F = GF(q)$. First, we study the zero divisors of $M_n(F)$. Let $a \in M_n(F)$, $a \notin F$ such that $L = F(a)$ is a field. Let l denote the degree of L over F. In other words, the minimal polynomial f of a over F is irreducible over F and $deg(f) = l$.

LEMMA 4.1. *There exists a* $c \in M_n(F)$ *such that:*

(i) $c^{-1}ac = a^q$;
(ii) *if* $Alg(a, c)$ *denotes the* F*-algebra generated by* a *and* c *then* $Alg(a, c)$ *is not commutative;*
(iii) $Alg(a, c) = L + cL + \cdots + c^m L + \cdots$.

PROOF. L is a simple subalgebra of $M_n(F)$ and the automorphism of L sending a to a^q leaves F element-wise fixed; therefore, by a theorem of Noether and Skolem (Pierce, 1982, section 12.6) this automorphism is inner, showing the existence of a $c \in M_n(F)$ satisfying (i). The statements (ii) and (iii) are valid for any such c. □

The next statement helps us to simplify the search for zero divisors. Let c be an arbitrary element satisfying (i) above.

LEMMA 4.2. *If* $Alg(a, c) = M_n(F)$ *then* $l = n$, $dim_F F(c) = n$, $c^n \in F$ *and*

$$Alg(a, c) = L \oplus cL \oplus \cdots \oplus c^{n-1} L. \tag{4.1}$$

PROOF. Simple calculation shows that for an arbitrary natural number i we have $c^{-i}ac^i = a^{q^i}$. This implies that $ac^l = c^l a$, therefore c^l is in the centre of $Alg(a, c) = M_n(F)$, which is F. The element c satisfies a polynomial of degree l over F, so we have

$$Alg(a, c) = L + cL + \cdots + c^{l-1} L,$$

by Lemma 4.1. This implies that $dim_F Alg(a, c) \leq l^2$. We have equality here iff the sum is a direct sum. As l is the degree of the minimal polynomial of a, we have $l \leq n$. Now from $dim_F Alg(a, c) = n^2$ we obtain that $l = n$ and the sum must be a direct sum. Finally, c cannot satisfy a polynomial over F with degree less than n for otherwise we had a decomposition shorter than (4.1), which is impossible. □

We continue the study of the case $Alg(a, c) = M_n(F)$. We have seen that c is a root of a polynomial of the form $x^n - \alpha$ where $\alpha \in F$. We know also that this is the minimal polynomial of c over F. The *norm* of an element d of L is defined as

$$norm(d) := dd^q d^{q^2} \cdots d^{q^{n-1}}.$$

(This is the L/F relative norm.) The next statement plays a key role in Herstein's proof of Wedderburn's theorem.

LEMMA 4.3. *Let $d \in L$ be an element such that $norm(d) = 1/\alpha$. Then $1 - cd$ is a zero divisor in the algebra $Alg(a, c) = M_n(F)$.*

PROOF. Let us define the element $z \in Alg(a, c)$ as

$$z = 1 + cd + c^2 dd^q + \cdots + c^{n-1} dd^q \cdots d^{q^{n-2}}.$$

A straightforward computation shows that $z(1 - cd) = 0$. On the other hand, the fact that (4.1) is a direct sum implies that neither z nor $1 - cd$ can be zero, proving the claim. □

We are unable to solve the above norm equation in general because of the high degree of the polynomial involved. To avoid this difficulty, we impose an additional restriction on c.

LEMMA 4.4. *Suppose that $Alg(a, c) = M_n(F)$ as above and that the minimal polynomial $x^n - \alpha$ of c is irreducible over F. Then $g(x) = x^n - 1/\alpha$ is also irreducible over F. Moreover, g splits into linear factors in L and if n is odd then $d \in L$, $g(d) = 0$ imply that $norm(d) = 1/\alpha$.*

PROOF. The irreducibility of $x^n - \alpha$ means that $F(c)$ is a field and $dim_F F(c) = n$. It is clear that $F(c) = F(1/c)$, therefore g is also irreducible over F. By comparing the degrees of the fields L and $F(c)$ we see that $F(c) \cong L$, so g splits into linear factors in L. As for the last statement, let d be an arbitrary root of g from L. The irreducibility of g implies that its constant term can be written as $(-1)^n norm(d) = -norm(d) = -1/\alpha$, giving the last statement. □

The proof of Lemma 4.4 covers the first part of the following lemma.

LEMMA 4.5. *Suppose that either n is odd or $n = 2$ and let L be an extension field of F such that $dim_F L = n$ and let $g(x) = x^n - \beta$, $\beta \in F$ be a given irreducible polynomial over F. Then we can find an element $d \in L$ such that $norm(d) = \beta$ by a polynomial time f-algorithm.*

PROOF. It suffices to consider the case $n = 2$. We distinguish two cases.

Case 1. $-\beta$ is a quadratic nonresidue in F. If $d \in L$ is a root of $h(x) = x^2 + \beta$ then the other root must be d^q and calculating the constant term of g we obtain that $d^{q+1} = norm(d) = \beta$.

Case 2. $-\beta$ is a quadratic residue in F. Let γ be an element of F such that $\gamma^2 = -\beta$. If we can find an element $b \in L$ such that $norm(b) = -1$ then by letting $d = \gamma b$ we obtain $norm(d) = \gamma^2 norm(b) = \beta$.

To solve the norm equation $norm(b) = -1$, it suffices to find a generating element b of the Sylow 2-subgroup of the multiplicative group L^* of L. Such an element b can be found by solving at most $2 \log_2 q$ quadratic equations in L. To prove that b is a solution, we remark that if b^{q+1} were a residue in F, then it would imply that

$$b^{(q-1)(q+1)/2} = 1,$$

contradicting the fact that b is a nonresidue in L. The element $norm(b)$ is in the Sylow 2-subgroup of F^*, therefore $norm(b) = -1$.

In all cases the norm equation can be solved by factoring moderately sized polynomials. The proof is complete. □

Now we are ready to describe our algorithm to find zero divisors in finite algebras. Suppose that we have an algebra A over the field $Z = GF(p^r)$, p prime and $dim_Z A = m$. The algebra A is given by structure constants. Our objective is to find a pair of zero divisors in A. The procedure ZERODIV( ) returns a pair of zero divisors x, y of A if there are any.

procedure ZERODIV(A)

Step 1. *Compute $Rad(A)$ using the method of section 2. If $Rad(A) \neq (0)$ then pick an arbitrary nonzero element $x \in Rad(A)$. As x is nilpotent, an appropriate power of it will suffice as y,* **return**(x, y).

Step 2. (**A is semisimple**)
Determine the Wedderburn decomposition of A, using the method of section 3. If A is not simple, say $A = I \oplus J$ where I and J are nonzero ideals, then x and y can be arbitrary nonzero elements of I and J, respectively.
return(x, y).

Step 3. (**A is simple**)
Check whether A is commutative. In case of an affirmative answer terminate concluding that A is a field (and therefore it has no zero divisors).

Step 4. (**A is a full matrix algebra over some field extension $F = GF(q)$ of Z, say $A \cong M_n(F)$ and $n > 1$.**)
Pick an arbitrary element $b \in A$ such that $b \notin F$. Next, compute and factor the minimal polynomial f of b over F. If f is reducible over F, say $f = gh$ a proper factorisation then **return**($g(b)$, $h(b)$).

Step 5. (**f is irreducible, therefore $F(b)$ is a field.**)
If $dim_F F(b)$ is even then choose an $a \in F(b)$ such that $dim_F F(a) = 2$ (we have to find a solution of the system of linear equations $a^{q^2} = a$ which is not in F), otherwise let $a := b$.

Step 6. (**$F(a)$ is a field and $l = dim_F F(a)$ is either odd or it is 2.**)
Find a nonzero element $c \in A$ such that $ac = ca^q$ (by solving a system of linear equations). Compute and factor the minimal polynomial of c over F. If it is reducible over F then **return** *zero divisors as in step 4.*

Step 7. (**c is an invertible element of A and $c^{-1}ac = a^q$ holds.**)
Form $Alg(a, c)$, the F-algebra generated by a and c. If $Alg(a, c) \neq A$ then let $A := Alg(a, c)$ and **go back to** *step 1.*

Step 8. (* *We have here* $n = l$, $Alg(a, c) = A \cong M_n(F)$, *n is either odd or* $n = 2$, *the minimal polynomial of c over F is* $f(a) = x^n - \alpha$ *for some* $\alpha \in F$ *and f is irreducible over F.**)
Find a solution $d \in F(a)$ *of the norm equation* $norm(d) = 1/\alpha$ *using the algorithm of Lemma* 4.5.
Now put $x := 1 - cd$ *and*

$$y := 1 + cd + c^2 dd^q + \cdots + c^{n-1} dd^q \cdots d^{q^{n-2}}.$$

return(x, y).

end procedure

THEOREM 4.6. *Let A be an algebra over* $GF(p^r)$, $dim_{GF(p^r)} A = m$. *The procedure ZERODIV( ) finds a pair of zero divisors in A if A contains zero divisors. As an f-algorithm it runs in time polynomial in m, r and* log *p*.

PROOF. The correctness and the timing of steps 1–3 is covered in Theorems 2.7 and 3.1. The procedure is essentially an iteration. If we enter the loop, i.e. we go back to step 1 from step 7, then *the actual algebra A is not commutative and thus contains zero divisors* by Lemma 4.1(ii). This statement may serve as a loop invariant. It is also clear that the dimension of A strictly decreases during an iteration step, except the last one. If we terminate at steps 4 or 6 then we clearly found zero divisors. The correctness of the annotation of step 8 follows from Lemma 4.2 and from the remark after it, so we can apply Lemma 4.3 to show that x and y form, indeed, a pair of zero divisors. We can solve the norm equation using the method of Lemma 4.5.

The structure of the loop shows that a step is executed at most m times. This implies a polynomial bound on the running time, provided that we have polynomial bounds on the time required by the individual steps. For steps 1–3 this follows from Theorems 2.7 and 3.1.

In steps 4–8 the major computational tasks are: solving systems of linear equations with at most m equations and at most m unknowns over F (or over Z); factoring polynomials of degree at most m over F; computing a (basis of a) subalgebra generated by two elements. These clearly can be done in time polynomial in the input size if we use an oracle for factoring polynomials. The norm equation from step 8 can be solved in polynomial time by Lemma 4.5 and the element y can then be computed efficiently using fast exponentiation. The proof is complete. □

For factoring polynomials over finite fields, we can use either the Las Vegas method of Berlekamp (1970), Rabin (1980) and Cantor & Zassenhaus (1981), or the deterministic method of Berlekamp (1968).

COROLLARY 4.7. *The problem of finding zero divisors in A can be solved by a Las Vegas method running in time polynomial in m, r and* log *p*. *Also, there exists a deterministic algorithm running in time polynomial in m, r and p.* □

5. Applications

We give three applications of the results to some algorithmic problems from computational algebra. Applications to special types of algebras will be considered elsewhere.

5.1. EXPLICIT ISOMORPHISMS OF MATRIX ALGEBRAS

From the results of sections 2 and 3 we can easily see that there exists an efficient f-algorithm to decide if a given finite algebra A is isomorphic to a full matrix algebra. Indeed, it suffices to check if $Rad(A) = (0)$ and whether A is directly indecomposible. In case of an affirmative answer, say if $A \cong M_n(F)$, then we can also find n and F. The problem remains to establish an explicit isomorphism from A to $M_n(F)$ (i.e. to represent A as an algebra of linear transformations of an n-dimensional linear space V over F). To find such a representation, it suffices to find an idempotent $e \in A$ such that $rank(e) = 1$ (here e is viewed as an element of $M_n(F)$; in the light of Lemma 5.1 the rank is independent from the actual isomorphism). Indeed, we can put $V = M_n(F)e$. It is well known that $dim_F V = n$ and $M_n(F)$ acts nontrivially and therefore faithfully on V via multiplication from the left.

The following easy lemma will be useful. The straightforward proof is omitted.

LEMMA 5.1. *Let $e \in M_n(F)$ be an idempotent such that $rank(e) = m$. Then $eM_n(F)e \cong M_m(F)$.*

This lemma shows that it is enough to give an algorithm to find a *singular* idempotent in A: if $e \in M_n(F)$, $rank(e) = m < n$ then we reduced the problem to the smaller instance $eAe \cong M_m(F)$. Indeed, if f is a rank one idempotent of eAe, then f is a rank one idempotent in A as well, because e is the identity element of eAe and this implies that $F \cong feAef = fAf$.

To obtain a singular idempotent, we call ZERODIV(A). If $n > 1$ then it returns a zero divisor x. Let e be the right identity element of the left ideal Ax. The element e is obviously a singular idempotent and can be found by solving a system of linear equations (describing that e is the identity element of Ax), once x is given. We have the following

THEOREM 5.2. *Suppose that we have an algebra A such that $A \cong M_n(GF(q))$. Then an explicit isomorphism from A to $M_n(GF(q))$ can be constructed by an f-algorithm running in time polynomial in n and $\log q$.* □

An explicit isomorphism is useful for the usual representation of $M_n(F)$ (as the algebra of all n by n matrices over F) is easy to handle. For example, we can conveniently decompose it into a direct sum of minimal left ideals. If e_{ii} denotes the matrix having 1 in position (i, i) and zeros elsewhere, then we have

$$M_n(F) = M_n(F)e_{11} \oplus M_n(F)e_{22} \oplus \cdots \oplus M_n(F)e_{nn}.$$

It is easy to check that $M_n(F)e_{ii}$ is a minimal left ideal $i = 1, \ldots, n$. If we have an explicit isomorphism from A to $M_n(F)$ then we can obtain a decomposition of A. This observation can be generalised to an arbitrary finite semisimple algebra B. Using the techniques of section 3, B can be decomposed into a direct sum of minimal ideals. The minimal ideals are simple algebras, hence we can decompose them into a direct sum of minimal left ideals using the above method. By putting these minimal left ideals together, we obtain a decomposition of B.

COROLLARY 5.3. *Let B be a semisimple algebra over $F = GF(q)$, $dim_F B = m$. B can be decomposed into a direct sum of minimal left ideals by an f-algorithm running in time polynomial in m and $\log q$.*

5.2. COMMON INVARIANT SUBSPACES

Now we apply our methods to solve an important linear algebraic problem over a finite ground field. Let $X_1, \ldots, X_k \in M_n(F)$ and consider their action on the linear space V of n by 1 column vectors over F. We would like to find a nontrivial subspace $U \leq V$ such that $X_i U \leq U$ for $i = 1, \ldots, k$, if there is any. This problem is solved by the procedure INV( ). It has one input parameter. It must be a nonempty set $S \subset M_n(F)$. For the sake of simplicity we also assume that the zero matrix is not in S. It outputs a proper invariant subspace if there is such a subspace.

procedure INV(S)

Step 1. *Compute A, the algebra of matrices generated by S (i.e. compute a basis of A over F).* (**S and A have the same invariant subspaces.**)

Step 2. *If $AV < V$ then* **return**(AV).

Step 3. *Compute Rad(A), the radical of A. If Rad(A) > (0) then* **return**(*Rad*(A)V).

Step 4. (**A is semisimple and V is a unitary A-module.**)
Break A into a direct sum of minimal left ideals:

$$A = \rho_1 \oplus \cdots \oplus \rho_m.$$

Let v be an arbitrary nonzero vector from V and consider the (A-invariant) subspaces $\rho_1 v, \ldots, \rho_m v$ and let U be any of these which is not (0). If $U = V$ then there is no nontrivial invariant subspace, otherwise **return**(U).

end procedure

First, we look at the correctness of INV( ). It is obvious that S and A have the same invariant subspaces. If we terminate at step 2 then we have a nontrivial invariant subspace because of the relation $AV > (0)$. If we terminate at step 3 then *Rad*(A)V is an A-invariant subspace. This subspace is different from (0), for $M_n(F)$ acts faithfully on V. As *Rad*(A) is a nilpotent algebra, the equality $V = Rad(A)V$ is impossible, hence we have a nontrivial invariant subspace. If we are at step 4 then A is semisimple. It is known (cf. Herstein, 1968, pp. 97–98) that $\rho_i v$ is either (0) or a minimal A-invariant subspace. The fact that we have survived the test of step 2 implies that V is a unitary A-module (i.e. the identity element of A is the identity matrix). This implies that not all of the subspaces $\rho_i v$ can be (0). We conclude that U is a minimal invariant subspace. In particular, if $U = V$, then V has no proper invariant subspaces. The correctness is proved.

Steps 1 and 2 can obviously be done in polynomial time. As for step 3, we have a polynomial time algorithm to compute *Rad*(A), cf. Theorem 2.7. By Corollary 5.3 the last step can also be done by a polynomial time f-algorithm. We have the following corollary.

COROLLARY 5.4. *Let $S = \{X_1, \ldots, X_k\} \subset M_n(F)$. A nontrivial invariant subspace can be found (if there is any) by an f-algorithm running in time polynomial in k, n and* $\log q$. *In particular, the problem can be solved by a Las Vegas method running in time polynomial in the parameters n, k,* $\log q$ *and by a deterministic method running in time polynomial in the parameters n, k, p and r, where $q = p^r$.* □

REMARK. The above methods can be extended to obtain an efficient method to solve a fundamental problem of representation theory: to decompose finite modules over finite semisimple algebras into a direct sum of simple submodules. In fact, if A is semisimple, then INV( ) returns a simple submodule of V and if we can find simple submodules, then we can find a direct decomposition using standard methods (cf. Herstein, 1968, pp. 97–98).

5.3. AN APPLICATION TO PERMUTATION GROUPS

In this section we give a deterministic polynomial time method to solve an algorithmic problem related to permutation groups; to find a minimal normal subgroup in an elementary abelian interval of permutation groups. More precisely, consider the following setting. Let $G \leq S_n$ and let $K < H$ normal subgroups of G such that H/K is an elementary abelian p-group for some prime p. Suppose that G, H, K are given by generating sets. Let m denote the number of generating elements we have. Our aim is to find a minimal normal subgroup L of G such that $K < L \leq H$.

First, we compute strong generating sets (cf. Luks, 1982) of the above groups. These have $O(n^2)$ elements. The factor group $V := H/K$ can be considered as a linear space of dimension $< n$ over $GF(p)$. The elements of G act on V as $GF(p)$-linear transformations via conjugation. Our problem is equivalent to finding a minimal G-invariant subspace of V. Obviously, it suffices to find a minimal Φ-invariant subspace, where Φ is the generating set of G we have. First, we compute a basis of V. This task can be done in polynomial time. Next, we compute matrix representations of the linear transformations corresponding to the elements of Φ. We can then apply INV( ) to find a minimal invariant subspace. As p is small, we can use the deterministic factoring method of Berlekamp (1968).

COROLLARY 5.4. *The above problem can be solved by a deterministic algorithm running in time polynomial in n and m.*

The author is indebted to L. Babai for suggesting the main problems considered here and for his constant interest and helpful comments. Helpful conversations on the subject with S. Becker, K. Friedl, W. M. Kantor and E. M. Luks are gratefully acknowledged. Thanks are due to the referee for helpful suggestions.

References

Babai, L. (1979). Monte-Carlo algorithms in graph isomorphism testing. *Tech. Rep. 79–10*, Department of Mathematics and Statistics, University of Montreal.

Berlekamp, E. R. (1968). *Algebraic Coding Theory*. New York: McGraw-Hill.

Berlekamp, E. R. (1970). Factoring polynomials over large finite fields. *Math. of Computation* **24,** 713–715.

Beck, R. E., Kolman, B., Stewart, I. N. (1977). Computing the structure of a Lie algebra. In: *Computers in Nonassociative Rings and Algebras*, pp. 167–188. London: Academic Press.

Ben-Or, M. (1981). Probabilistic algorithms over finite fields. *Proc. 22th IEEE FOCS*, 394–398.

Camion, P. (1983). A deterministic algorithm for factorizing polynomials of $F_q[x]$. *Annals of Discrete Math.* **17,** 149–157.

Cantor, D. G., Zassenhaus, H. (1981). A new algorithm for factoring polynomials over finite fields. *Math. of Computation* **36,** 587–592.

Dickson, L. E. (1923). *Algebras and Their Arithmetics*. Chicago, IL: University of Chicago Press.

Friedl, K. (1983). *Algorithms in Algebra* (in Hungarian). Diploma Thesis. Eötvös University, Budapest.

Friedl, K., Rónyai, L. (1985). Polynomial time solutions of some problems in computational algebra. *Proc. 17th ACM STOC,* pp. 153–162. Providence, RI.

Herstein, I. N. (1968). *Noncommutative Rings*. Mathematical Association of America.

Humphreys, J. E. (1980). *Introduction to Lie algebra and representation theory*. *GTM* **9.** New York: Springer.

Hopcroft, J. E., Ullman, J. D. (1979). *Introduction to Automata Theory, Languages and Computation*. Reading, MA: Addison-Wesley.

Jacobson, N. (1962). *Lie Algebras*. New York: John Wiley.

Knuth, D. E. (1981). *The Art of Computer Programming, Vol. 2. Seminumerical Algorithms*. Reading, MA: Addison-Wesley.

Kertész, A. (1987). *Lectures on Artinian Rings* (ed. R. Wiegandt). Budapest: Akadémiai Kiadó.

Lidl, R., Niederreiter, H. (1983). *Finite Fields*. Reading, MA: Addison-Wesley.

Luks, E. M. (1982). Isomorphism of graphs of bounded valence can be tested in polynomial time. *J. Comput. Syst. Sci.* **25,** 42–65.

Pierce, R. S. (1982). *Associative Algebras*. Berlin: Springer.

Rabin, M. O. (1980). Probabilistic algorithms in finite fields. *SIAM J. Comp.* **9,** 273–280.

Interpolating Polynomials from their Values

RICHARD ZIPPEL
Department of Computer Science, Cornell University
Ithaca, New York 14853 *U. S. A.*

(*Received* 4 *September* 1989)

A fundamental technique used by many algorithms in computer algebra is interpolating polynomials from their values. This paper discusses two algorithms for solving this problem for sparse multivariate polynomials, an updated version of a probabilistic one and a new deterministic technique that uses some ideas due to Ben-Or and Tiwari (1988). In addition algorithms are presented for quickly finding points that are not zeroes of sparse multivariate polynomials—the zero avoidance problem.

1. Introduction

Mathematical calculations involving polynomials or other symbolic quantities suffer from a problem not found in numerical calculations: *intermediate expression swell*. That is, when performed in a straightforward fashion, the intermediate expressions of a calculation are much larger than the final answer. Fundamentally, this difference is due to the fact that the amount of space required to represent the product of two floating point numbers is about as much as for each of the original multiplicands. However, the space required for the product of two multivariate polynomials can be much larger than that required for the multiplicands. In fact, even the sum of two multivariate polynomials can be twice as large the summands. This effect is more pronounced with polynomials with many variables.

Two fundamental approaches to this problem have been suggested. Each generates one or more simplified computations where some of the symbolic variables are replaced by numerical values. These simplified problems do not suffer as much from intermediate expression swell and may be solved more easily than the original problem. The two techniques differ in how they determine the solution of the original problem from the solutions of the simplified ones.

The first approach, which we call the *modular technique*, solves a large number of simplified problems but uses carefully chosen values for the symbolic variables. These solutions are then interpolated to recover the variables eliminated in the simplified problems, producing the final answer. In many practical algorithms the resulting intermediate expressions do not involve any symbolic variables and there is essentially no intermediate expression swell. This interpolation technique was first introduced in the modular GCD algorithm of Brown (1971).

The second approach, which we call *Newton's technique*, uses the solution to a single simplified problem as a the initial value for a p-adic solution derived by Newton's iteration. (Conversion of a p-adic solution to a solution in the original ring is rarely difficult.) This is the basic idea behind the polynomial factoring algorithm of Wang and Rothschild (1975), the EZGCD algorithm of Moses and Yun (1973) and its successors Wang (1978) and most of the polynomial factoring algorithms now in use. Both the modular technique and Newton's technique suffer when the answer is sparse (has relatively few non-zero coefficients). In this case a great deal of effort is expended computing coefficients that are zero.

Versions of both the modular technique and Newton's technique whose time complexity is random polynomial were first given in Zippel (1979, 1980). Applications of these techniques to polynomial factoring and their analysis and extension, have been presented in a number of papers by von zur Gathen and Kaltofen: von zur Gathen (1983, 1985), Kaltofen (1985a, 1985b, 1987) and von zur Gathen and Kaltofen (1985). The probabilistic nature of these algorithms stems from an assumption about certain polynomials that arise in the calculation. It is known that the values of these polynomials at certain points are zero. This could happen either if the polynomials were identically zero or if the points chosen happened to be zeros of the polynomials. The key assumption of these algorithms is that the polynomials are identically zero. These algorithms can be made deterministic by choosing points that cannot all be zeroes of these polynomials. We call this the *zero avoidance problem*.

Problem. (Zero Avoidance Problem) *Given some set of parameters for a polynomial (number of variables, degree, number of non-zero terms, size of coefficients, etc.) choose a set of points $\mathcal{S}$ such that no polynomial with those parameters vanishes at all of the points of $\mathcal{S}$.*

The original sparse polynomial algorithms used only the number of variables (n) and degree (d) parameters in choosing the set $\mathcal{S}$. It is easy to show that $\mathcal{S}$ must contain at least $(d+1)^n$ points (see proposition 1 in section 2). To prove that a polynomial is zero using this set of points would require time exponential in the number of variables. Thus fast algorithms that use only these parameters are probabilistic. The deterministic algorithms given here also make use of the number of non-zero terms (T) in choosing $\mathcal{S}$. It is this additional bit of information about the polynomial that keeps the size of $\mathcal{S}$ small.

Many of the ideas used to solve the zero avoidance problem can be used to clarify and simplify certain steps in the modular technique. The particular piece that we discuss in this paper we call the *interpolation problem*. Rather than choosing points to prove that a polynomial is not identically zero, we go further and actually determine the polynomial itself.

Problem. (Interpolation Problem) *Given a set of parameters for a polynomial (number of variables, degree, number of non-zero terms, size of coefficients, etc.) choose a set of points $\mathcal{S}$ with the following property. For any polynomial P with those parameters, P can be determined from $\mathcal{S}$ and $P(\mathcal{S})$ quickly, i.e. either polynomial time or probabilistic polynomial time.*

In this paper we present three solutions to the zero avoidance problem, and two solutions to the interpolation problem. Each is summarized in the following two tables.

Zero Avoidance Problem			
	Schwartz	Zippel	Ben-Or Tiwari
Algorithm type	Probabilistic	Deterministic	Deterministic
Number of evaluations	1	nT^2	$T+1$
Chance of error	ϵ	0	0
Size of evaluations in bits	$\log \frac{nd}{\epsilon}$	$nT \log T$	$T \log n$

The column labeled "Schwartz" corresponds to the probabilistic algorithm presented by J. Schwartz (1980) and which is intrinsic to Zippel (1979). Since it does not take into consideration the number of non-zero terms in the P, the parameter T does not appear. In the third column, labeled "Ben-Or Tiwari," we give the recent results of Ben-Or and Tiwari (1988). The second column, labeled "Zippel," is a new algorithm presented here in section 5. Though its performance is inferior to that of Ben-Or and Tiwari it makes use of some new techniques that may be of use in other problems. In particular, it yields a deterministic solution of a variant of the zero avoidance problem for polynomials over finite fields.

For the interpolation problem a new parameter arises, t the true number of non-zeroes terms in P. This can be much smaller than the *a priori* bound on the number of non-zero terms T.

The first column of this table characterizes the author's original probabilistic algorithm updated to include an idea of Ben-Or and Tiwari. The third column corresponds to the deterministic algorithm due to Ben-Or and Tiwari (1988). It is unique in that it does not require *a priori* bounds on the degrees of the variables that appear in the result. Notice that the probabilistic algorithm is significantly better than the deterministic one when the bound on the number of terms is not sharp ($T \gg t$). The second "Zippel" algorithm is a new deterministic variant

Interpolation Problem			
	Zippel	Zippel	Ben-Or Tiwari
Algorithm type	Probabilistic	Deterministic	Deterministic
Degree bounds	Yes	Yes	No
Number of operations	ndt^2	ndt^2T	$T^2(\log^2 T + \log nd)$
Number of evaluations	ndt	$ndtT$	$2T$
Size of evaluations in bits	$\log \frac{nd^2T^2}{\epsilon}$	$T \log n$	$T \log n$

of the probabilistic algorithm whose dependence on T is not quite so strong as Ben-Or and Tiwari's algorithm. Thus it also performs especially well when T is not a sharp bound. This algorithm is presented in section 6.

Kaltofen and Yagati (1988) have suggested an improved technique for solving the systems of linear equations that arise in the two interpolation algorithms discussed in this paper. Their ideas improve the algorithms discussed in the paper to give the performance figures given above. In this table $M(t)$ denotes the complexity of multiplying two univariate polynomials of degree t. This variant of the deterministic algorithm is competitive with Ben-Or and Tiwari's algorithm.

Interpolation Problem		
	Kaltofen-Yagati	Kaltofen-Yagati
Algorithm type	Probabilistic	Deterministic
Degree bounds	Yes	Yes
Number of operations	$ndM(t)\log t$	$ndTM(t)\log t$
Number of evaluations	ndt	$ndtT$
Size of evaluations in bits	$\log \frac{nd^2T^2}{\epsilon}$	$T \log\log n$

In the conclusions we give some comments on how these algorithm impact some of the original calculations, such as greatest common divisor and factorization problems.

2. Generalities

We let $\mathbf{Z}$ denote the rational integers and $\mathbf{Z}/(m)$ the integers modulo m. $\mathbf{F}_q$ denotes the finite field with q elements and $\mathbf{F}_q^*$ its multiplicative subgroup.

Throughout this paper we assume polynomials are represented as a list of monomials (pairs of exponent vectors and coefficients) and that monomials with zero coefficients are omitted. The number of variables in a polynomial is denoted by n. Thus the exponent vectors are n-tuples. The maximum degree of any variable in the polynomial is denoted by d. The number of non-zero monomials of the polynomial P is usually denoted by t or terms(P), for additional preciseness. For a *dense polynomial*, one where each monomial has a non-zero coefficient, terms$(P) = (d+1)^n$. We generally use capital letters to denote *a priori* bounds, and lower case letters for the actual value. Thus T is used to designate a bound on the number of terms in P, while t denotes the actual number of non-zero terms present in P.

To minimize the number of subscripts in formulae we use a variant of the notation introduced by Laurent Schwartz. Let $\vec{X} = (X_1, X_2, \ldots, X_n)$ and $\vec{e} = (e_1, e_2, \ldots, e_n)$ be two vectors. Then we write the usual (inner) dot product as

$$\vec{e} \cdot \vec{X} = e_1 X_1 + e_2 X_2 + \cdots + e_n X_n.$$

We also extend this notation to exponentiation as follows

$$X^{\vec{e}} = (X^{e_1}, X^{e_2}, \ldots, X^{e_n}) \qquad \text{and} \qquad \vec{X}^{\vec{e}} = X_1^{e_1} X_2^{e_2} \cdots X_n^{e_n}.$$

Thus the multivariate polynomial

$$c_1 X_1^{e_{11}} X_2^{e_{12}} \cdots X_n^{e_{1n}} + c_2 X_1^{e_{21}} X_2^{e_{22}} \cdots X_n^{e_{2n}} + \cdots + c_t X_1^{e_{t1}} X_2^{e_{t2}} \cdots X_n^{e_{tn}}$$

would be written

$$c_1 \vec{X}^{\vec{e}_1} + c_2 \vec{X}^{\vec{e}_2} + \cdots + c_t \vec{X}^{\vec{e}_t}.$$

We always use the vector accent when using this notation.

When evaluating algorithms involving polynomials, we need to measure the size of a polynomial. In this paper we have chosen to use the number of non-zero terms. Thus an upper bound on the size of a polynomial of n variables, each of degree d, is $(d+1)^n$. The number of non-zero terms, however, is often much smaller. Notice that when establishing that a polynomial P of size $O(T)$ is identically zero, we already know that P cannot have more than $O(T)$ non-zero coefficients, though we know little about the exponents.

An alternative measure of the size of a polynomial P is the size of a straight line program to compute P. This measure was advanced by Kaltofen (1987). The class of straight line programs of size $O(T)$ contains almost all polynomials with $O(T)$ non-zero terms and many more. It would be interesting to know if it is possible to extend the results presented here to this wider class of polynomials.

To prove that a polynomial is zero by considering its value at a number of points requires some bound on the information content of the polynomial. We begin with a proposition that establishes a lower bound for our results.

Proposition 1. *Let $\mathcal{S} = \{\vec{a}_i\}$ be a set of $T-1$ n-tuples. There exists a polynomial with rational integer coefficients, not identically zero, that contains no more than T non-zero monomials and that vanishes at every point in $\mathcal{S}$.*

Proof: Choose a polynomial with T monomials:

$$P(\vec{X}) = c_1 \vec{X}^{\vec{e}_1} + c_2 \vec{X}^{\vec{e}_2} + \cdots + c_T \vec{X}^{\vec{e}_T},$$

whose coefficients (c_i) will be determined later from $\mathcal{S}$, $\vec{e}_i \neq \vec{e}_j$ and chosen arbitrarily. For P to vanish at $\vec{a}_i$, an element of $\mathcal{S}$, the c_j must satisfy the following linear equation:

$$c_1 \vec{a}_i^{\vec{e}_1} + c_2 \vec{a}_i^{\vec{e}_2} + \cdots + c_T \vec{a}_i^{\vec{e}_T} = 0.$$

Since these equations are homogeneous and there are more undetermined variables than linear constraints, there is a non-trivial solution to this system of equations. □

Assume we wish to prove that a polynomial is zero using only its value at points that we are free to specify. Proposition 1 demonstrates that showing the polynomial is zero at T points only shows that that the polynomial is either identically zero or has more than T non-zero terms. Thus if all that is known about a polynomial is the number of variables, n and degree bounds on those variables, d, we will need $(d+1)^n$ evaluations to prove that the polynomial is non-zero. This means that there is no deterministic algorithm that proves a polynomial is zero from its values, and degree bounds. Additional information is also needed.

For univariate polynomials over the reals, we can show that by choosing the points carefully, any polynomial with no more than T terms that vanishes at T points is identically zero.

Proposition 2. *Let $P(x)$ be a univariate polynomial with coefficients in* $\mathbf{Z}$. *The number of positive real zeroes of $P(x)$ is less than or equal to* $\mathrm{terms}(P) - 1$.

Proposition 2 follows immediately from Descartes' rule of signs since the maximum number of sign changes in the coefficients of $P(x)$ is $\mathrm{terms}(P) - 1$. (For instance, Pólya and Szegö (1976) Part V, Chapter 1, problem 36.) The following corollary is merely a restatement of the proposition.

Corollary. *A univariate polynomial that vanishes at the integers $1, 2, \ldots, T$ is either identically zero or has more than T non-zero coefficients.*

Using some new techniques we show in section 5 that $O(nT^2)$ points suffice, where n is the number of variables in P. This is accomplished by finding a specialization of P to one variable that does not increase the number of non-zero terms and then applying Proposition 2. The previous best results were that $(d+1)^n + 1$ sufficed, which is optimal for dense polynomials, but can be exponentially bad for sparse polynomials. In section 6 we give Ben Or and Tiwari's result that T suffices. In light of proposition 1 this is the best possible.

We occasionally use the notation p_i to indicate the ith prime. It is is also used to represent the ith element of the vector $\vec{p}$. Our intent should be clear from the context. Later, we will need a crude estimate for the size of the product of the first N primes. For our work we can use the crude estimate of

$$p_N \leq (N+1)^{1+\epsilon},$$

for some small constant ϵ. This is much weaker than the best known results, for instance Rosser and Schoenfeld (1962).

By applying Sterling's formula to the product of the first N primes we have

$$\begin{aligned}\log(p_1 p_2 \cdots p_N) &= \log(N+1)!^{1+\epsilon} \\ &= (1+\epsilon)\left(N+\frac{3}{2}\right)\log(N+2) + O(N)\end{aligned}$$

This proves the following proposition:

Proposition 3. *There exists a constant c_1 such that*

$$\log(p_1 p_2 \cdots p_N) \leq c_1 N \log N.$$

Many of the algorithms developed in this paper depend upon the special properties of Vandermonde matrices, which we summarize here. A *Vandermonde matrix* is a matrix of the form

$$V_n = \begin{pmatrix} 1 & k_1 & k_1^2 & \cdots & k_1^{n-1} \\ 1 & k_2 & k_2^2 & \cdots & k_2^{n-1} \\ \vdots & & \vdots & & \vdots \\ 1 & k_n & k_n^2 & \cdots & k_n^{n-1} \end{pmatrix},$$

where the k_i are chosen from some field. Similarly, a system of linear equations of the form

$$\begin{aligned} X_1 + k_1 X_2 + k_1^2 X_3 + \cdots + k_1^{n-1} X_n &= w_1 \\ X_1 + k_2 X_2 + k_2^2 X_3 + \cdots + k_2^{n-1} X_n &= w_2 \\ &\vdots \\ X_1 + k_n X_2 + k_n^2 X_3 + \cdots + k_n^{n-1} X_n &= w_n \end{aligned}$$

will be called a *Vandermonde system of equations*.

A matrix where the degrees of each row rise monotonically, but not necessarily linearly, is called a *generalized Vandermonde matrix*, viz,

$$\begin{pmatrix} k_1^{e_1} & k_1^{e_2} & k_1^{e_3} & \cdots & k_1^{e_n} \\ k_2^{e_1} & k_2^{e_2} & k_2^{e_3} & \cdots & k_2^{e_n} \\ \vdots & & \vdots & & \vdots \\ k_n^{e_1} & k_n^{e_2} & k_n^{e_3} & \cdots & k_n^{e_n} \end{pmatrix}.$$

The following well known theorem gives the determinant of a Vandermonde matrix.

Proposition 4. *The determinant of the Vandermonde matrix is*

$$\det V_n = \prod_{1 \le i < j \le n} (k_j - k_i).$$

As an immediate consequence the determinant of a Vandermonde matrix is non-zero if and only if the k_i are distinct.

A similar result is true for generalized Vandermonde matrices over the reals, but the proof is a little trickier. Notice that while proposition 4 applied to Vandermonde matrices over any field, the following proposition is only valid over the reals, which has characteristic zero. We know of no similar result for fields of positive characteristic.

Proposition 5. *The determinant of a generalized Vandermonde matrix is non-zero if the k_i are distinct positive real numbers.*

A proof of this result can be found in Gantmacher (1959), volume II, page 99.

The inverse of a Vandermonde matrix can be computed by the following well known technique. (See Press (1986), for example.) Multiply a Vandermonde matrix by a general n by n matrix:

$$\begin{pmatrix} 1 & k_1 & k_1^2 & \cdots & k_1^{n-1} \\ 1 & k_2 & k_2^2 & \cdots & k_2^{n-1} \\ \vdots & & \vdots & & \vdots \\ 1 & k_n & k_n^2 & \cdots & k_n^{n-1} \end{pmatrix} \cdot \begin{pmatrix} a_{11} & a_{12} & a_{13} & \cdots & a_{1n} \\ a_{21} & a_{22} & a_{23} & \cdots & a_{2n} \\ \vdots & & \vdots & & \vdots \\ a_{n1} & a_{n2} & a_{n3} & \cdots & a_{nn} \end{pmatrix}$$

The jth element of the top row of the product of these two matrices is

$$a_{1j} + a_{2j}k_1 + a_{3j}k_1^2 + \cdots + a_{nj}k_1^{n-1} = P_j(k_1).$$

In fact the product above is

$$\begin{pmatrix} P_1(k_1) & P_2(k_1) & \cdots & P_n(k_1) \\ P_1(k_2) & P_2(k_2) & \cdots & P_n(k_2) \\ \vdots & \vdots & & \vdots \\ P_1(k_n) & P_2(k_n) & \cdots & P_n(k_n) \end{pmatrix}.$$

Choosing the $P_j(Z)$ to be

$$P_j(Z) = \prod_{\substack{i \ne j \\ 1 \le i \le n}} \frac{Z - k_i}{k_j - k_i},$$

we see that the product matrix is the identity, and thus the coefficients of the P_j are the columns of the inverse of the Vandermonde matrix. Each of the $P_j(Z)$ can be computed in $O(n)$ operations from a master polynomial, which itself can be computed in $O(n^2)$ operations. Thus the Vandermonde matrix can be inverted in $O(n^2)$ time.

Assume we wish to solve a Vandermonde system of equations like the following:

$$\begin{aligned} X_1 + k_1X_2 + k_1^2X_3 + \cdots + k_1^{n-1}X_n &= w_1 \\ X_1 + k_2X_2 + k_2^2X_3 + \cdots + k_2^{n-1}X_n &= w_2 \\ &\vdots \\ X_1 + k_nX_2 + k_n^2X_3 + \cdots + k_n^{n-1}X_n &= w_n \end{aligned} \tag{1}$$

If recognized as a Vandermonde system, these equations need only consume $O(n)$ space. They can be solved using $O(n)$ space by the following device.

Define

$$P(Z) = \prod_{1 \le i \le n} (Z - k_i).$$

This polynomial contains $n + 1$ terms. The coefficient of Z^n is always 1. The polynomials $P(Z)/(Z - k_j)$ can be computed by synthetic division. It is the numerator of $P_j(Z)$. The value of $P(Z)/(Z - k_j)$ at k_j is the denominator of $P_j(Z)$. Thus each of the $P_j(Z)$ can be computed using $O(n)$ space and time. The computation of the X_i is arranged as follows.

$$\begin{pmatrix} X_1 \\ \vdots \\ X_n \end{pmatrix} = \begin{pmatrix} w_1 \cdot \mathrm{coef}(P_1, Z^0) \\ \cdots \\ w_1 \cdot \mathrm{coef}(P_1, Z^{n-1}) \end{pmatrix} + \cdots + \begin{pmatrix} w_n \cdot \mathrm{coef}(P_n, Z^0) \\ \cdots \\ w_n \cdot \mathrm{coef}(P_n, Z^{n-1}) \end{pmatrix}.$$

After each column vector on the right hand side is computed, it is added to the accumulating X_i and its storage may be reused by the following column vector.

This approach can also be applied to transposed Vandermonde systems like following

$$\begin{aligned} X_1 + X_2 + X_3 + \cdots + X_n &= w_1 \\ k_1X_1 + k_2X_2 + k_3X_3 + \cdots + k_nX_n &= w_2 \\ &\vdots \\ k_1^{n-1}X_1 + k_2^{n-1}X_2 + k_3^{n-1}X_3 + \cdots + k_n^{n-1}X_n &= w_2 \end{aligned} \tag{2}$$

since the inverse of the transpose of a matrix is the transpose of the inverse, we have the following formula for each of the X_i

$$X_i = w_1 \cdot \mathrm{coef}(P_1, Z^0) + w_2 \cdot \mathrm{coef}(P_1, Z^1) + \cdots + w_n \cdot \mathrm{coef}(P_1, Z^{n-1}).$$

These results are summarized in the following proposition.

Proposition 6. *The Vandermonde system (1) and the transposed Vandermonde system (2) over the field F can be solved in $O(n^2)$ operations over F. Furthermore, the space required is that of $O(n)$ elements of F. If $F = \mathbf{Q}$ and $K = \max|\operatorname{num} k_i| + \max|\operatorname{den} k_i|$ then the largest number used will require $O(n \log K)$ bits and in total $O(n^2 \log K)$ bits of storage will be required.*

3. Dense Interpolation

The general problem we consider in this paper is computing a polynomial from its values at certain points, whose choice may be part of some higher level algorithm. These polynomials may be multivariate and their coefficients generally lie in the rational integers, though occasionally they lie in a finite field. Many of these results can be extended to more complex fields, but we do not do this here. In this section we assume the number of variables in the polynomial is given, as well as degree bounds, but no additional assumptions are made. In particular, nothing is known about the number of non-zero terms in the polynomial.

3.1 Univariate Dense Interpolation

The simplest form of this problem consists of determining a univariate polynomial from its values at selected points. The straight forward approach works surprisingly well. Let

$$P(Z) = p_0 + p_1 Z + \cdots + p_{n-1} Z^{n-1} + p_n Z^n$$

be the polynomial to be determined. Assume the coefficients are over a field F, and let $z_0, \ldots, z_n$ be the set of distinct evaluation points. From the values of $P(z_i) = w_0$ we get the following system of linear equations in the unknown p_j.

$$\begin{aligned}
p_0 + p_1 z_0 + p_2 z_0^2 + \cdots + p_n z_0^n &= w_0 \\
p_0 + p_1 z_1 + p_2 z_1^2 + \cdots + p_n z_1^n &= w_1 \\
&\vdots \\
p_0 + p_1 z_n + p_2 z_n^2 + \cdots + p_n z_n^n &= w_n
\end{aligned}$$

This is a Vandermonde system and can solved quickly using the algorithm of Proposition 6.

3.2 Multivariate Dense Interpolation

As pointed out in the previous section, a polynomial in one variable of degree d can be determined from its values at $d+1$ points using $O(d^2)$ arithmetic operations. This result can be extended to multivariate problems.

Let $P(\vec{X})$ be the polynomial to be determined. It is a polynomial in n variables, $X_1, \ldots, X_n$, whose coefficients lie in an integral domain R. Each X_i appears to degree no more than d_i in P. Let $\ell = (d_1+1)(d_2+1)\cdots$, the maximum number of terms in P. Writing P as a sum of monomials using the vector notation we have

$$P(\vec{X}) = c_1\vec{X}^{\vec{e}_1} + c_2\vec{X}^{\vec{e}_2} + \cdots + c_\ell\vec{X}^{\vec{e}_\ell},$$

where the $\vec{e}_i$ run through each possible exponent combination. Choosing ℓ random n-tuples $\vec{x}_i$ and computing the values of $P(\vec{x}_i)$ gives a system of ℓ linear equations. In general, this requires $O(\ell^3)$ operations to solve, and perhaps more important $O(\ell^2)$ space.

There remains the question of solvability of the system of equations. Let M be an $n \times n$ matrix over a field F. M will be singular if and only if $\det M = 0$. Thus the singular matrices form an algebraic set of codimension 1 in the space of all $n \times n$ matrices. Thus the probability that the system of equations is singular is about $1/\#(F)$. For probabilistic algorithms this suffices. For deterministic algorithms more analysis is required. This is done in later sections by choosing the evaluation points carefully.

A recursive technique was used by Brown (1971) in the modular GCD algorithm was first used to bring the time requirements for interpolation down to $O(\ell^2)$. In this paper we use another approach that leads more naturally to the techniques for dealing with sparse polynomials.

Choose a random n-tuple $\vec{p}$. This is the initial evaluation point. Denote the values of the monomials $\vec{p}^{\vec{e}_i}$ by m_i. Additional evaluation points are obtained by raising $\vec{p}$ to successive powers (starting with 0). Notice that $\left(\vec{p}^j\right)^{\vec{e}_j} = m_i^j$. Thus we have the following system of equations to solve.

$$\begin{aligned}
c_1 + c_2 + \cdots + c_\ell &= P(\vec{p}^0) \\
c_1 m_1 + c_2 m_2 + \cdots + c_\ell m_\ell &= P(\vec{p}) \\
c_1 m_1^2 + c_2 m_2^2 + \cdots + c_\ell m_\ell^2 &= P(\vec{p}^2) \\
&\vdots \\
c_1 m_1^{\ell-1} + c_2 m_2^{\ell-1} + \cdots + c_\ell m_\ell^{\ell-1} &= P(\vec{p}^{\ell-1})
\end{aligned}$$

This is the transpose of a Vandermonde system. As discussed in section 2, this system can be solved in $O(\ell^2)$ time and $O(\ell)$ space.

The key issue in this approach is guaranteeing the m_i are distinct so that the Vandermonde system will be non-singular. If the coefficient domain, R, is a unique factorization domain we can do this easily. For instance, assume R is the rational integers. We choose the components of $\vec{p}$ to be distinct primes, viz., $\vec{p} = (2, 3, 5, \ldots)$. By unique factorization each of the m_i will be distinct.

If the coefficient domain is a finite field $\mathbf{F}_q$, then the problem can be more difficult. The finite field must contain at least ℓ elements for the Vandermonde system to be non-singular. For the dense interpolation technique being discussed, the maximum value of ℓ is $(d+1)^n$. When $q < (d+1)^n$ the modular interpolation technique can still be used but elements should be chosen from an algebraic extension of $\mathbf{F}_q$ that has more than $(d+1)^n$ elements. In general, we can solve the system of equations using conventional $(O(\ell^3))$ techniques.

If the characteristic of $\mathbf{F}_q$ is sufficiently large, we can do better. Choose the components of $\vec{p}$ to be the rational primes, $(2, 3, 5, \ldots)$. If each of the m_i, when computed in $\mathbf{Z}$, is less than the characteristic of $\mathbf{F}_q$ then they will be distinct as elements of $\mathbf{F}_q$. For this to be the case the characteristic must be greater than

$$(2 \cdot 3 \cdots p_n)^d \approx n^{c_1 nd},$$

by proposition 3.

This idea of substituting primes for each of the variables was first suggested by Grigoriev and Karpinksi (1987), who were studying a problem involving polynomials with 0/1 coefficients. These ideas were first applied to interpolation by Tiwari (1987).

In the following paragraphs we analyze the hard case: We assume that q is greater than $(d+1)^n$, but that the characteristic of $\mathbf{F}_q$ is less than $n^{c_1 nd}$. The actual analysis is staightforward but somewhat lengthy. We consider the following somewhat more general question since its solution will be of use in analyzing the sparse algorithms. Let $\{\vec{e}_i\}$ be a set of T n-tuples where each component is bounded by d. (In the current case $T = (d+1)^n$.) What is the probability that for a randomly chosen $\vec{x} \in \mathbf{F}_q^n$ there is an $\vec{e}_i$ and $\vec{e}_j$ such that $\vec{x}^{\vec{e}_i} = \vec{x}^{\vec{e}_j}$?

We begin with an elementary enumeration proposition. The one dimensional version can be found in almost any book on elementary number theory.

Proposition 7. *Let $\vec{a} \neq 0$ be a fixed n-tuple where each component is an element of $\mathbf{Z}/(m)$ and c be the common GCD of the a_i and m. Let $\vec{x}$ be an n-tuple whose components range over $\mathbf{Z}/(m)$. Then $\vec{a} \cdot \vec{x}$ takes on m/c distinct values. These values divide the different $\vec{x}$ into m/c classes each containing cm^{n-1} different $\vec{x}$.*

Proof: First we reduce to the case where $c = 1$. Since $\vec{a} \cdot \vec{x}$ is a multiple of c for every $\vec{x}$, $\vec{a} \cdot \vec{x}$ can take on no more than m/c values, i.e. $0, c, 2c, \ldots$. Let αc be one of these values. Each solution of

$$\frac{\vec{a} \cdot \vec{x}}{c} \equiv \alpha \pmod{m/c} \tag{3}$$

gives rise to c^n solutions of $\vec{a} \cdot \vec{x} \equiv c\alpha \pmod{m}$. Thus if we can show that (3) has $(m/c)^{n-1}$ solutions, we are finished. The rest of the proof proceeds via a slightly complicated induction.

Consider the one dimensional case, $ax \equiv b \pmod{m}$. Since a and m are relatively prime, there is exactly one value of x that satisfies the relation for every value of b, as required by the proposition.

Now assume the proposition is true for all vectors $\vec{a}$ of dimension less than n. Let α be an arbitrary element of $\mathbf{Z}/(m)$. We want to show that $\vec{a} \cdot \vec{x} \equiv \alpha \pmod{m}$ has m^{n-1} zeroes. Without loss of generality we can assume that a_1 is not zero. If a_1 and m are relatively prime then for every choice of $a_2, \ldots, a_n$ there is a unique a_1 that satisfies the relation. Thus there are m^{n-1} zeroes of the relation as desired.

Assume that a_1 and m have a GCD of g. The relation has no zeroes if g does not divide $a_2x_2 + \cdots a_nx_n - \alpha$. Thus we consider the number of zeroes of

$$a_2x_2 + \cdots a_nx_n \equiv \alpha \pmod{g}.$$

Notice that $a_2, \ldots, a_n$ cannot have a GCD dividing g. Thus this equation has g^{n-2} zeroes modulo g. Each is the image of $(m/g)^{n-1}$ elements modulo m. Thus there are m^{n-1}/g choices of $a_2, \ldots, a_n$. Each one will give rise to g choices for x_1 giving the desired result. □

Corollary. *Let $\vec{a}$, m and c be as in the previous proposition. Then there are cm^{n-1} distinct solutions to $\vec{a} \cdot \vec{x} = 0 \pmod{m}$.*

Proof: 0 is always one of the values of $\vec{a} \cdot \vec{x}$ since $\vec{x}$'s components could be all zeroes. □

This result can now be used to answer the question raised above.

Proposition 8. *Let $\vec{e}_1, \ldots, \vec{e}_T$ be n-tuples where each component is less than d. There exists no more than*

$$\frac{d \cdot T \cdot (T-1) \cdot (q-1)^{n-1}}{2}$$

n-tuples $\vec{X}$ with components in F_q such that for some i and j $\vec{X}^{\vec{e}_i}$ and $\vec{X}^{\vec{e}_j}$ have the same values. Equivalently, for at least

$$(q-1)^{n-1}\left(q - 1 - \frac{d \cdot T \cdot (T-1)}{2}\right)$$

n-tuples $\vec{X}$, $\vec{X}^{\vec{e}_i}$ takes on distinct values.

Proof: Let g be a generator of the multiplicative group $\mathbf{F}_q^*$. Then for each n-tuple $\vec{X}$ we can assign another n-tuple $\vec{a}$ such that $X_i = g^{a_i}$, assuming no X_i is zero. The a_i are elements of $\mathbf{Z}/(q-1)$. Two monomials $\vec{X}^{\vec{e}_i}$ and $\vec{X}^{\vec{e}_j}$ have the same value when

$$\vec{x}^{\vec{e}_i} = g^{\vec{a} \cdot \vec{e}_i} = g^{\vec{a} \cdot \vec{e}_j} = vecx^{\vec{e}_j}.$$

That is, when $\vec{a} \cdot (\vec{e}_i - \vec{e}_j) = 0 \pmod{q-1}$. By the previous corollary there are $c(q-1)^{n-1}$ such zeroes, where c is the GCD of the elements of $\vec{e}_i - \vec{e}_j$ and $q-1$. Since $c \leq d$ there are at most $d(q-1)^{n-1}$ tuples $\vec{x}$ that cause these two terms to take on the same value.

There are $T(T-1)/2$ distinct pairs of $\vec{e}_i$, so the maximum number of $\vec{x}$ that cause a pair of $\vec{x}^{\vec{e}_i}$ to take on the same value is

$$\frac{d \cdot T \cdot (T-1) \cdot (q-1)^{n-1}}{2}.$$

□

Since there are only $(q-1)^{n-1}$ possible $\vec{x}$ (ignoring those with a zero component), we have the following corollary.

Corollary. *The probability that a randomly chosen $\vec{x}$ will cause two of the $\vec{x}^{\vec{e}_i}$ to have the same value is*

$$\frac{d \cdot T \cdot (T-1)}{2(q-1)}.$$

If we wish the probability of a collision to be less than ϵ, then for dense polynomials this means that

$$q > \frac{(d+1)^{2v+1}}{2\epsilon}.$$

This is actually quite impractical for polynomials with large numbers of variables and high degree. Fortunately, many problems are sparse, i.e. $T \ll (d+1)^v$, which gives much better results. This is the topic of the next section.

4. Sparse Interpolation

The purpose of this section is to develop Zippel's sparse interpolation algorithm, which gives a probabilistic resolution of the interpolation problem. What is presented is an improvement of the author's original results based on some of the ideas first suggested by Ben-Or and Tiwari. This algorithm is given no information about the number of non-zero terms in the polynomial being interpolated. Instead it develops an estimate of the number of terms as each new variable is introduced. As a consequence its performance depends upon the actual number of non-zero terms in the polynomial rather than an *a priori* bound. This probabilistic algorithm tends to be more useful in practical situations than the deterministic algorithms presented in the following sections.

This section has been divided into three subsections. In the first we give a demonstration of the algorithm and its benefits. In the subsequent subsections we give a more formal presentation of its details, and analyze the algorithm's performance.

4.1 Heuristic Presentation

As before we wish to determine the polynomial $P(\vec{X}) \in U[\vec{X}]$ from its values, where U is a field with sufficiently many distinct elements. We assume that d_i bounds the degree of X_i in P. The sparse interpolation algorithm computes P one variable at a time. That is, we initially compute $P(a_1, a_2, \ldots, a_n)$, then $P(X_1, a_2, \ldots, a_n)$, then $P(X_1, X_2, a_3, \ldots, a_n)$ and so on, until we have determined $P(\vec{X})$. The introduction of each new variable is called a *stage* of the algorithm. We use clues from the polynomial produced in the preceding stage to minimize the effort required to produce the next polynomial in the sequence.

The description of the sparse interpolation algorithm becomes rather involved and it is easy to get bogged down by all the subscripts and variables involved, but it is fundamentally quite simple. In this section we give an explicit example.

Assume we wish to interpolate a polynomial in three variables, $P(X, Y, Z)$ over a field, where the degree of each variable is not greater than 5. When the polynomial is dense, there are 125 different coefficients that must be determined. We assume that most of these coefficients are zero and that P possesses only a few non-zero monomials. By using one of the dense interpolation schemes of section 3, we can compute $P(X, y_0, z_0)$ from $P(x_0, y_0, z_0), P(x_1, y_0, z_0), \ldots, P(x_5, y_0, z_0)$. Assume this yields

$$P(X, y_0, z_0) = c_0 X^5 + c_1 X + c_2.$$

This is the end of the first stage.

Beginning stage two, we know that $P(X, Y, Z)$ can be written as

$$P(X, Y, Z) = P_5(Y, Z) X^5 + P_4(Y, Z) X^4 + \cdots + P_0(Y, Z).$$

From the first interpolation we know that $P_5(y_0, z_0) = c_0$, $P_1(y_0, z_0) = c_1$ and $P_0(y_0, z_0) = c_2$. Since the other coefficients are zero

$$P_4(y_0, z_0) = P_3(y_0, z_0) = P_2(y_0, z_0) = 0.$$

The key step in the sparse interpolation algorithm is to assume that this is true for all values of Y and Z. That is, that

$$P_4(Y, Z) = P_3(Y, Z) = P_2(Y, Z) = 0.$$

In typical calculations, where y_0 and z_0 are chosen at random from a large set of possibilities, this is a good assumption. Proposition 9 below gives a precise measure of how good an assumption this is.

We now choose a new value for Y, y_1, and compute $P(X, y_1, z_0)$. Without the assumption of the previous paragraph, this interpolation would require 6 additional values of P. Instead we assume that $P(X, y_1, z_0)$ contains only 3 non-zero terms, i.e.,

$$P(X, y_1, z_0) = c_3 X^5 + c_4 X + c_5,$$

where the c_3, c_4 and c_5 are to be determined. Since there are only three unknowns to be determined only three new values of P are required.

This process is repeated until we have six polynomials.

$$\begin{aligned} c_0 X^5 + c_1 X + c_2 &= P(X, y_0, z_0) \\ c_3 X^5 + c_4 X + c_5 &= P(X, y_1, z_0) \\ &\vdots \\ c_{15} X^5 + c_{16} X + c_{17} &= P(X, y_5, z_0) \end{aligned}$$

By the dense interpolation algorithm of section 3, the coefficients of the X^5 terms can be interpolated to produce a polynomial in Y, and similarly for the linear and constant terms. Combining these results we have $P(X, Y, z_0)$. Notice that we have only needed $6 + 3 + 3 + 3 + 3 + 3$ values of P to compute this polynomial. The dense interpolation scheme would have required almost twice as many evaluation points.

Beginning the third stage, let us assume this gives the polynomial

$$\begin{aligned} P(X, Y, z_0) &= k_1 X^5 + (k_2 Y^4 + k_3 Y)X + k_4 Y^5 \\ &= k_1 X^5 + k_2 X Y^4 + k_3 X Y + k_4 Y^5, \end{aligned}$$

where the k_i are elements of the ground field. We are now in a position to begin the process again, but this time introducing the variable Z. To do this we need to calculate the polynomials $P(X, Y, z_0), P(X, Y, z_1), \ldots, P(X, Y, z_5)$. We assume that those XY-monomials that did not arise in $P(X, Y, z_0)$ have coefficients of zero in $P(X, Y, Z)$.

Thus to compute

$$P(X, Y, z_1) = k_5 X^5 + k_6 X Y^4 + k_7 X Y + k_8 Y^5$$

we only need interpolate four values of P. Thus the additional 5 polynomials only required $5 \times 4 = 20$ evaluation points. Without the sparsity assumptions each of the 5 polynomial would have required 36 evaluation points, and 180 in all.

4.2 Formal Presentation

To fix our notation, assume we want to use the sparse interpolation algorithm to determine a polynomial $P(X_1, \ldots, X_n) \in F[\vec{X}]$ where we know that each X_i does not appear to degree higher than d and that there are t non-zero monomials in P. Furthermore, we assume that we can compute the value of P given a value for $\vec{X}$. It is convenient to consider just one stage of the interpolation. The computation of $P(\vec{X})$ being just a sequence of n stages.

Now assume that we have performed the first $k-1$ stages of the sparse modular algorithm and we are about to begin the kth stage. From the previous stage's computation we have

$$P(X_1, \ldots, X_{k-1}, x_{k0}, \ldots, x_{n0}) = p_{10}\vec{X}^{\vec{e}_1} + p_{20}\vec{X}^{\vec{e}_2} + \cdots + p_{T0}\vec{X}^{\vec{e}_T}.$$

The set of exponents of $P(X_1, \ldots, X_{k-1}, x_{k0}, \ldots, x_{n0})$ is called the *skeleton* of P, which we denote by $\operatorname{skel} P$. Since there are t non-zero monomials in P, the skeleton of P can never have more than t elements.

Throughout this stage, the values assigned to $X_{k+1}, \ldots, X_n$ do not vary. To simplify the notation, we will omit them.[1] We write

$$P'(y_1, \ldots, y_k) = P(y_1, \ldots, y_k, x_{k+1,0}, \ldots, x_{n0}).$$

The computation of $P(X_1, \ldots, X_{k-1}, X_k)$ proceeds in two phases. In the first we determine

$$P'(X_1, \ldots, X_{k-1}, x_{kj}) = p_{1j}\vec{X}^{\vec{e}_1} + p_{2j}\vec{X}^{\vec{e}_2} + \cdots + p_{Tj}\vec{X}^{\vec{e}_T},$$

for $j = 0, \ldots, d$ by the following technique:

For each of $d+1$ randomly chosen values of X_k, x_{kj} perform the following. Pick a random $k-1$-tuple denoted by $(y_1, \ldots, y_{k-1}) = \vec{y}$, such that each of $\vec{y}^{\vec{e}_i}$ are distinct. Since the $\vec{e}_i$ are known, verifying that this is the case is easy. Actually finding $\vec{y}$ is discussed in the analysis of the next subsection. This value $\vec{y}$ allows us to set up the following (non-singular) system of linear equations

$$\begin{aligned}
P'(1, \ldots, 1, x_{kj}) &= p_{1j} + p_{2j} + \cdots + p_{Tj} \\
P'(y_1, \ldots, y_{k-1}, x_{kj}) &= p_{1j}\vec{y}^{\vec{e}_1} + p_{2j}\vec{y}^{\vec{e}_2} + \cdots + p_{Tj}\vec{y}^{\vec{e}_T} \\
P'(y_1^2, \ldots, y_{k-1}^2, x_{kj}) &= p_{1j}\vec{y}^{2\vec{e}_1} + p_{2j}\vec{y}^{2\vec{e}_2} + \cdots + p_{Tj}\vec{y}^{2\vec{e}_T} \\
&\vdots \\
P'(y_1^T, \ldots, y_{k-1}^T, x_{kj}) &= p_{1j}\vec{y}^{t\vec{e}_1} + p_{2j}\vec{y}^{t\vec{e}_2} + \cdots + p_{tj}\vec{y}^{t\vec{e}_t}
\end{aligned}$$

This is a Vandermonde system of equations and can be solved by the techniques of Section 2 in $O(t^2)$ time while requiring only $O(t)$ space. The result will be a polynomial

$$P(X_1, \ldots, X_{k-1}, x_{kj}, x_{k+1,0}, \ldots, x_{n0}),$$

[1] In practical implementations this may be more than notational. Eliminating the variables that do not vary at this stage can save significant time when computing the values of P.

for each of the $d+1$ values x_{kj}.

In the second phase, we independently interpolate the coefficients of each monomial, using the dense interpolation algorithm. The results of these interpolations can be combined to produce

$$P'(X_1,\ldots,X_{k-1},X_k) = p_1(X_k)\vec{X}^{\vec{e}_1} + p_2(X_k)\vec{X}^{\vec{e}_2} + \cdots + p_T(X_k)\vec{X}^{\vec{e}_T}.$$

The dense interpolation yielded the univariate polynomials $p_i(X_k)$. This polynomial is in turn expanded to get

$$P(X_1,\ldots,X_k,x_{k+1,0},\ldots,x_{n0}) = p'_{10}\vec{X}^{\vec{e}'_1} + p'_{20}\vec{X}^{\vec{e}'_2} + \cdots + p'_{T0}\vec{X}^{\vec{e}'_T},$$

and we are ready to begin lifting the next variable.

4.3 Analysis

We begin by presenting the probabilistic resolution of the zero avoidance problem. The following proposition gives a sharp estimate of how difficult it is to avoid the zeroes of a polynomial given only degree bounds.

Proposition 9. *Let k be a field, f any element of $k[X_1,\ldots,X_n]$ such that the degree of X_i in f is bounded by d_i. Let $Z_n(B)$ be the number of zeroes of f, $\vec{x}$ such that $x_i \in \mathcal{S}$ (a set with B elements $B \gg d_i$). Then*

$$\begin{aligned} Z_n(B) &\le B^n - (B-d_1)(B-d_2)\cdots(B-d_n) \\ &\approx O\left((d_1+d_2+\cdots+d_n)B^{n-1}\right). \end{aligned}$$

Historical note: This result initially appeared in two papers simultaneously and independently at the EUROSAM '79 conference in Marseille during the summer of 1979. Schwartz gave the second estimate of this proposition while Zippel gave a version of the first. The proof given here is a simplification and extension of that given in Zippel (1979).

Proof: There are at most d_n values of X_n at which f is identically zero. So for any of these d_n values of X_n and any value for the other X_i, f is zero. This comes to d_nB^{n-1}. For all other $B-d_n$ values of X_n we have a polynomial in $n-1$ variables. The polynomial can have no more than $Z_{n-1}(B)$ zeroes. Therefore,

$$Z_n(B) \le d_nB^{n-1} + (B-d_n)Z_{n-1}(B).$$

Rather than solving this recurrence for Z_n, we solve it for $N_n = B^n - Z_n$. Since Z_1 is less than or equal to d_1, $N_1 \ge (B-d_1)$. This is the basis step of the inductive proof. Writing the recurrence in terms of N_n we have

$$B^n - N_n(B) \le d_nB^{n-1} + (B-d_n)(B^{n-1} - N_{n-1}(B))$$

or

$$N_n(B) \geq (B - d_n)N_{n-1}(B),$$

from which the proposition follows. □

Polynomials of the following actually have $B^n - (B - d_1) \cdots (B - d_n)$ zeroes with components less than B

$$f(X_1, \ldots, X_n) = \prod_{0 \leq i_1 \leq d_1} (X_1 - x_{1i_1}) \quad \cdots \quad \prod_{0 \leq i_n \leq d_n} (X_n - x_{ni_n}).$$

Thus the inequality in the proposition cannot be further strengthened. The following corollary phrases this result as a probability.

Corollary. *Let $f_1, f_2, \ldots, f_s$ be elements of $k[X_1, \ldots, X_n]$, where the degree in each variable is bounded by d. Let $\mathcal{P}(f_1, \ldots, f_s)$ be the probability that a randomly chosen point $\vec{x}$ is a zero of any of the f_i, where x_i is an element of a set with B elements. Then*

$$\mathcal{P}(f_1, \ldots, f_s) < \frac{nds}{B}.$$

Proof: Let $f = f_1 f_2 \cdots f_s$. The degree of each variable in f is bounded by ds. Applying the previous proposition, we see that the number of zeroes of f is bounded by $ndsB^{n-1}$, for sufficiently large B. Since there are B^n possible $\vec{x}$ to choose from, we have the corollary. □

This corollary gives the probabilistic solution to the zero avoidance problem. Let P be an element of $\mathbf{Z}[X_1, \ldots, X_n]$. Choose a random point in $\mathbf{Z}^n$ with components less than B. The probability that this point will be a zero of P is less than

$$\frac{nd}{B}.$$

Thus to keep the chance of error below ϵ, using a single evaluation, we must choose

$$B > \frac{nd}{\epsilon}.$$

Turning now to the sparse modular interpolation algorithm, if all the probabilistic assumptions hold, the cost of lifting a single variable can be computed as follows. In phase 1 we compute $d+1$ polynomials at a cost of at most t evaluation points each, requiring $O(dt^2)$ time and $O(dt)$ space. The dense interpolation in phase 2 requires $O(d^2)$ steps for each coefficient that is interpolated. At most t such interpolations are performed so a total of $O(td^2)$ steps are required. Since n stages need to be performed the total time requirements are $O(ndt(d+t))$, while the maximum space requirement is always $O(dt)$. Remember that t is the actual number of terms in P, not an *a priori* bound on the number of terms in P.

As we shall see, the chance for error in the interpolation depends entirely on the initial evaluation point $(x_{10}, x_{20}, \ldots, x_{n0})$. By performing several interpolations with different initial points we can decrease the chance of error. This may be appropriate in practical implementations. Here we use ϵ to denote the chance for error from a single starting point. We assume P is a polynomial over a finite field, $\mathbf{F}_q$. We want to determine ϵ as a function of q, d, n and t. In practical implementation, P will most likely be a polynomial over $\mathbf{Z}$. Then q is chosen to minimize ϵ and still remain efficient. To convert from a solution in a finite field to one over the integers a coefficient bound is needed for the solution in $\mathbf{Z}$. In this section we ignore these issues. From a theoretical point of view, we continue to compute in $\mathbf{Z}$ (or $\mathbf{Q}$ as necessary), and restrict our random choices to integers less than q.

There are two sources of potential error in this algorithm. First, the structure inherited from earlier stages in phase 1 structure may be incorrect. That is, a term that was assumed to be zero really wasn't zero. To be precise, consider a three variable problem. Assume the polynomial to be computed is

$$p_1(Y,Z)X^{e_1} + p_2(Y,Z)X^{e_2} + \cdots + p_t(Y,Z)X^{e_t},$$

and the initial evaluation point is (x_0, y_0, z_0). After the single variable interpolation computed in stage one we have the polynomial

$$p_1(y_0,z_0)X^{e_1} + p_2(y_0,z_0)X^{e_2} + \cdots + p_t(y_0,z_0)X^{e_t}.$$

In passing from the one variable case (X) to the two variable case (X, Y), the algorithm just presented assumes the structure given above is correct. If p_i vanished at (y_0, z_0) we would have assumed e_i was zero erroneously. At the end of this stage we will have the bivariate polynomial

$$q_1(z_0)(X,Y)^{\vec{f}_1} + q_1(z_0)(X,Y)^{\vec{f}_2} + \cdots + q_t(z_0)(X,Y)^{\vec{f}_t}.$$

Again, if any of the q_i vanish at z_0 we will get erroneous results. To compute the exponent vectors correctly, we need to assume that the p_i and q_i do not vanish at the initial point (x_0, y_0, z_0). These are the polynomials whose zeroes we must avoid.

At the ith stage of the interpolation process, there are at most t polynomials in $n-i$ variables whose zeroes must be avoided. The aggregate number of non-zero terms these polynomials contain must be less than t. The degrees of each polynomial is bounded by d. So by proposition 9, the chance of the initial evaluation point being the zero of any of these is

$$\frac{n^2 dt}{q}.$$

Now consider the probability that the Vandermonde systems are singular. These systems are of rank at most t. By the corollary to proposition 8, the probability that this system is singular is

$$\frac{d \cdot t \cdot (t-1)}{2(q-1)} \approx \frac{dt^2}{2q}.$$

At each stage there are d such systems to be solved and there are $n-1$ stages in all, so the probability that one of them will fail is bounded by

$$\frac{nd^2t^2}{q}.$$

Thus we have the following proposition.

Proposition 10. *Let P be a polynomial in n variables, each of degree no more than d and with t ($\gg n$) non-zero terms. Assume the coefficients of P lie in a finite field with q elements. The probability that the sparse interpolation algorithm will give the wrong answer for this polynomial is less than*

$$\frac{nd^2t^2}{q}.$$

The randomly chosen values must be chosen from a set of at least

$$\frac{nd^2t^2}{\epsilon}$$

values for the probability of error to be less than ϵ.

Since we cannot know the true number of non-zero terms of P before beginning the algorithm, the random values must be chosen from a set of

$$\frac{nd^2T^2}{\epsilon}$$

points.

5. Deterministic Zero Avoidance

As mentioned earlier, proposition 2 shows that univariate polynomials over the rational integers cannot have many real zeroes. We extend this proposition to one for a multivariate polynomial in the variables X_i by finding a substitution ($X_i \mapsto Z^{e_i}$) that sends a multivariate polynomial into a univariate polynomial. We then apply the proposition to the univariate polynomial to get our result. The crucial part of the proof is to show that we can find a substitution such that $P(Z^{\vec{e}})$ is not identically zero.

Before, proceeding with our version of the bounds, it is instructive to examine the bound derivable from Kronecker's technique, van der Waerden (1953). We are given a polynomial in n variables, $P(X_1, \ldots, X_n)$ where the maximum degree in any one variable is d and assume there are no more than T non-zero terms in P. Let ℓ be an integer larger than d. Consider the substitution, $X_i \mapsto Z^{\ell^{i-1}}$. A monomial $\vec{X}^{\vec{e}}$ is mapped to a monomial in Z raised to the power:

$$e_1 + e_2\ell + e_3\ell^2 + \cdots + e_n\ell^{n-1}.$$

Since each of the e_i are strictly smaller than ℓ this mapping is one to one and invertible.

Furthermore, we haven't changed the number of non-zero terms in the polynomial, i.e. terms($P(\vec{X})$) = terms($P(Z)$). By proposition 1, if $P(Z)$ vanishes for T positive values of Z, then it (and thus $P(\vec{X})$) is identically zero. This would be our desired proposition if the values chosen for the X_i were small enough. The smallest integer values we can choose for Z are $1, 2, \ldots, T$. Thus the values for X_i are

$$1^{\ell^{i-1}}, 2^{\ell^{i-1}}, \ldots, T^{\ell^{i-1}}.$$

Unfortunately, the size of the largest substitution, $T^{\ell^{n-1}}$ is exponential in the number of variables.

This basic idea can be salvaged by a more flexible choice of exponent substitutions. Rather than using an invertible substitution, as Kronecker does, we choose one that merely guarantees that $P(Z)$ is not identically zero if $P(\vec{X})$ is not. In light of the results of Ben-Or and Tiwari the importance of this result is somewhat diminished. However the technique used to reduce a multivariate polynomial to a univariate polynomial, while preserving the number of non-zero terms seems quite powerful.

We begin with a definition and some lemmas.

Definition: 1. Let $\mathcal{A}$ be a set of n-tuples with components in a ring R. $\mathcal{A}$ is said to be *maximally independent* if every subset of n elements of $\mathcal{A}$ is R-linearly independent.

In our situation, each element of $\mathcal{A}$, $\vec{s}$, corresponds to a substitution $X_i \mapsto Z^{s_i}$. The following lemma shows that there exist sets of N maximally independent n-tuples with entries not much larger than N.

Lemma 1. *Let S be a positive integer, and p the smallest prime larger than S. There exists a maximally independent set of S n-tuples with components in* $\mathbf{Z}$ *where each of the components of the n-tuples is less than p.*

Proof: First we show that we can construct arbitrarily large maximally independent sets of n-tuples. Then by reducing them modulo a prime we get the n-tuples required by the lemma. Consider n-tuples of the form $(1, k, k^2, \ldots, k^{n-1})$. For n of these to be independent the determinant of the matrix

$$\begin{pmatrix} 1 & k_1 & k_1^2 & \cdots & k_1^{n-1} \\ 1 & k_2 & k_2^2 & \cdots & k_2^{n-1} \\ \vdots & & \vdots & & \vdots \\ 1 & k_n & k_n^2 & \cdots & k_n^{n-1} \end{pmatrix}$$

must not be zero. Since this is a Vandermonde matrix, its determinant is $\prod_{i>j}(k_i - k_j)$, by proposition 3.

Thus if the k_i are distinct the vectors they generate will be linearly independent. In particular if we let $\vec{u}_k = (1, k, \ldots, k^{n-1})$ then any subset of n of the $\vec{u}_k$ will be linearly independent. Furthermore, if we reduce the elements of $\vec{u}_k$ by a prime larger than any of the k, the n-tuples remain maximally independent. □

Lemma 2. *Let $L(\vec{X}) = \vec{w} \cdot \vec{X}$ be a linear form in the n variables X_i that is not identically zero. If $\vec{p}_1, \ldots, \vec{p}_n$ are linearly independent n-vectors, then $L(\vec{p}_i) \neq 0$ for some i.*

Proof: Since the n-tuples $\vec{p}_i = (p_{i1}, p_{i2}, \ldots, p_{in})$ are linearly independent, the matrix

$$\mathbf{A} = \begin{pmatrix} p_{11} & p_{12} & \cdots & p_{1n} \\ p_{21} & p_{22} & \cdots & p_{2n} \\ \vdots & & & \vdots \\ p_{n1} & p_{n2} & \cdots & p_{nn} \end{pmatrix}$$

is non-singular. Denote by $\vec{w}$ the column vector $(w_1, \ldots, w_n)^T$. If L vanishes at each of the n-tuples $\vec{p}_i$ then

$$\mathbf{A} \cdot \vec{w} = 0.$$

Since $\mathbf{A}$ is non-singular, $\vec{w}$ must be identically zero. □

Lemma 3. *Let $L_j(\vec{X}) = \vec{w}_j \cdot \vec{X}$ be a set of T linear forms in n variables X_i, where none of the forms is identically zero. There exists a set of $(n-1) \cdot T + 1$ n-tuples such that for one of these n-tuples none of the L_j vanish. Furthermore, the components of these n-tuples can be chosen such that each component is less than $2nT$.*

Proof: By the previous lemma, each L_i can vanish at no more than $n-1$ independent n-tuples. Assuming none of the forms vanish at the same n-tuple, there can only $(n-1) \cdot T$ n-tuples for which one of the forms vanish. □

This Lemma can be extended somewhat to give a estimate of the number of n-tuples required to ensure that each linear form takes on a distinct value. This is important enough to justify calling it a proposition.

Proposition 11. *Let $L_j(\vec{X}) = \vec{w}_j \cdot \vec{X}$ be a set of T distinct linear forms in n variables X_i. There exists a set of*

$$\frac{(n-1)\cdot T\cdot(T-1)}{2}+1$$

n-tuples such that each L_j takes on a different value for at least one of them, and where the components are each less than $nT(T-1)$.

Proof: Consider the set of forms

$$M_{ij} = (\vec{w}_i - \vec{w}_j)\cdot\vec{X}.$$

Ignoring the diagonal forms (M_{ii}), which are identically zero, there are $T(T-1)/2$ distinct forms up to sign. L_i and L_j have the same value for some n-tuple, if and only if M_{ij} vanishes at the same n-tple. By Lemma 3, there exists a set of

$$\frac{(n-1)\cdot T\cdot(T-1)}{2}+1$$

n-tuples such that for one them, none of the M_{ij} vanish, and each has components less than $nT(T-1)$. □

Proposition 12. *There exists a set of nT^2 n-tuples such that there is no polynomial with less than T non-zero terms that vanishes at each of the n-tuples. Furthermore, the absolute value of the components of the n-tuples is less than T^{2nT}, and they have size $O(nT\log T)$.*

This proposition is proven by applying the same type of reasoning used earlier with Kronecker's trick, using a sufficiently large, maximally independent set of n-tuples.

Proof: Assume $P(X_1,\ldots,X_n)$ is not identically zero and let the terms of P be

$$P(\vec{X}) = c_1\vec{X}^{\vec{e}_1} + c_2\vec{X}^{\vec{e}_2} + \cdots + c_T\vec{X}^{\vec{e}_T}.$$

The substitution $X_i \mapsto Z^{u_i}$ sends this polynomial into

$$P(Z) = c_1 Z^{\vec{e}_1\cdot\vec{u}} + c_2 Z^{\vec{e}_2\cdot\vec{u}} + \cdots + c_T Z^{\vec{e}_T\cdot\vec{u}}.$$

This substitution must be chosen so that $P(Z)$ is not identically zero. This can be done by requiring that for any $i \neq 1$

$$\vec{e}_1\cdot\vec{u} \neq \vec{e}_i\cdot\vec{u}$$

or equivalently $(\vec{e}_i - \vec{e}_1)\cdot\vec{u} \neq 0$.

By Lemma 3, we can choose a set of $(n-1)(T-1)+1$ maximally independent n-tuples such that one of them satisfies $(\vec{e}_i - \vec{e}_1) \cdot \vec{u} \neq 0$. We can bound the components of $\vec{u}$ by p where p be the smallest prime larger than $(n-1)(T-1)+1$. Notice that $p < 2nT$.

Each of the n-tuples gives rise to a mapping from $P(\vec{X})$ to $P(Z)$. Since $P(Z)$ has no more than T monomials, we need not try more than T positive integer values for each $\vec{u}$. In particular we can use the values $1, 2, \ldots, T$ for Z. Thus there exists a set of $(n-1)(T-1)T + T < nT^2$ points that satisfies the requirements of the theorem. Furthermore, each component of the substitution is bounded by $T^p < T^{2nT}$. □

Proposition 13. *There exists a set of nT^2 n-tuples, whose components are of size $O(nT \log T)$, such that for every set of polynomials $P_i \in \mathbf{Z}[x_1, \ldots, x_n]$ with*

$$\sum_i \text{terms}(P_i) < T,$$

there is at least one n-tuple where none of the polynomials vanish.

This proposition follows from proposition 12 and the observation that if $t_1 + \cdots + t_k = T$, then the maximum value of $t_1^2 + \cdots + t_n^2$ is T^2.

The remaining result in this section is due to Ben-Or and Tiwari (1988). By using a direct multivariate approach to the zero-avoidance problem, they improve the $O(nT^2)$ result of proposition 12 for the zero avoidance problem to T, which is best possible. Ben-Or and Tiwari's main idea for this problem is contained in the following proposition.

Proposition 14. *Let $P(\vec{X})$ be a non-zero polynomial in $R[\vec{X}]$ with at most T terms and with monomial exponent vectors $\vec{e}_i$. Assume there exists an n-tuple $\vec{x}$ such that the $\vec{x}^{\vec{e}_i}$ are distinct. Then not all of $P(\vec{x}^0), P(\vec{x}), P(\vec{x}^2), \ldots, P(\vec{x}^{T-1})$ are zero.*

Proof: Denote $\vec{x}^{\vec{e}_i}$ by m_i. By assumption, each of the m_i are distinct. If P vanished at each of the $\vec{x}^i$ then the following system of linear equations would hold.

$$\begin{aligned} c_1 + c_2 + \cdots c_T &= 0 \\ c_1 m_1 + c_2 m_2 + \cdots + c_T m_T &= 0 \\ c_1 m_1^2 + c_2 m_2^2 + \cdots + c_T m_T^2 &= 0 \\ &\vdots \\ c_1 m_1^{T-1} + c_2 m_2^{T-1} + \cdots + c_T m_T^{T-1} &= 0 \end{aligned}$$

Since this is a Vandermonde system and we have assumed that the m_i are distinct, the system of equations is non-singular. Thus the c_i must all be zero, and P must be identically zero for all of $P(\vec{x}), P(\vec{x}^2), P(\vec{x}^3), \ldots, P(\vec{x}^{T-1})$ to vanish. □

The key then is finding a substitution that keeps the monomials distinct. If P is a polynomial over a unique factorization domain (such as the rational integers) then this is relatively easy—we choose the components of $\vec{x}$ to be distinct primes. In this case each of the m_i must be distinct by unique factorization. For polynomials over finite fields estimates of the difficulty in finding such the right initial substitution can be made form proposition 7, but this leads to a probabilistic algorithm.

The following proposition considers the zero avoidance problem for several polynomials.

Proposition 15. *Let $P_1(\vec{X}), \ldots, P_s(\vec{X})$ be non-zero polynomials in $U[X_1, \ldots X_n]$, U a unique factorization domain and assume that* $\operatorname{terms} P_1 + \cdots + \operatorname{terms} P_s = T$. *Let $\vec{x}$ be a vector of n primes in U. Then for integer j, $0 \le j \le T$, all of $P_i(\vec{x}^j)$ are different from zero.*

Proof: Denote the points $\{\vec{x}^0, \vec{x}, \ldots, \vec{x}^T\}$ by $\mathcal{A}$. P_i cannot vanish at more than $\operatorname{terms} P_i$ elements of $\mathcal{A}$ by proposition 5. Since $\mathcal{A}$ contains $T+1$ points, there must be one for which none of the P_i vanish.

6. New Interpolation Algorithm

Using either of the deterministic solutions of the zero avoidance problem given in the previous section (propositions 13 and 15), it is possible to modify the probabilistic sparse polynomial interpolation algorithm of section 4 to make it deterministic.

As usual, we wish to interpolate a sparse polynomial with no more than T non-zero terms, $P(\vec{X}) \in F[X_1, \ldots, X_n]$, from its values. As in the last section we will only consider the case when F is the rational integers or a finite field of sufficiently large characteristic. For simplicity our discussion will use $F = \mathbf{Z}$. Thus we can guarantee that the Vandermonde systems of equations are always non-singular, by using as the initial starting point: $(2, 3, 5, \ldots, p_n)$, where p_n is the nth prime.

The only remaining source of erroneous answers in the probabilistic algorithm of section 4, is that coefficient polynomials may vanish at the starting point. To be more precise, assume the starting point of the interpolation is $x_{10}, x_{20}, \ldots, x_{n0}$. Consider stage k, where we are introducing X_k. We can write $P(\vec{X})$ as

$$P(\vec{X}) = p_{1k}(X_{k+1}, \ldots, X_n)(X_1, \ldots, X_k)^{\vec{e}_1} + \cdots \\ + p_{tk}(X_{k+1}, \ldots, X_n)(X_1, \ldots, X_k)^{\vec{e}_t}.$$

If the polynomials p_{ik} do not vanish at the starting point, then skeleton produced at stage k will be a correct image of $\operatorname{skel} P$. If this is the case we say the starting point is a *stage k good starting point*. If the starting point was not good, then the resulting skeleton will be strictly smaller than the correct one at that stage.

The deterministic version of the sparse modular algorithm assumes that at stage $k-1$, the polynomial it is given has the correct skeleton. It then produces a k variable polynomial that has the correct skeleton, by ensuring that it has used a starting point for which none of the p_{ik} vanish. This is easily done by performing the operations of stage k, T times, using $(x_{k+1,0}^{s}, \ldots, x_{n0}^{s})$ as the values for the undetermined variables. Since the total number of terms in p_{ik} not greater than T, by proposition 14, one of these starting points will be stage k good. Since we know the correct $k-1$ skeleton it is not necessary to repeat lower stages of the algorithm.

Thus this algorithm will require T times more operations than the probabilistic version. The components of the evaluation points are always primes (or a random integer for X_n). Thus the largest component will be p_n^T, whose size is approximately $O(T \log n)$.

7. Conclusions

We have presented new deterministic solutions to both the zero avoidance problem and the interpolation problem for sparse polynomials. The zero avoidance technique of proposition 13 reduces multivariate problems to univariate problems. The interpolation algorithm presented may have better performance than Ben-Or and Tiwari's interpolation algorithm if the bound on the number of terms is not sharp.

Unfortunately, these deterministic results do not immediately yield deterministic algorithms for the multivariate polynomial greatest common divisor (GCD) and factorization problems. For the GCD problem a technique for avoiding the zeroes of the resultant of the two polynomials is needed. Unfortunately, straight forward estimates of the number of terms of the resultant are exponential in the number of variables even if the original polynomials were sparse. For the factorization problem, using the current techniques, there still remains the need for an effective version of the Hilbert Irreducibility theorem with good constants. The existing versions give probabilistic results, von zur Gathen (1983), Heintz and Sieveking (1981) and Kaltofen (1985b).

8. Acknowledgements

This work benefited greatly by discussions with Barry Trager. The one dimensional version of these ideas, Proposition 2, was suggested by Gregory Chudnovsky. The key idea in Lemma 1 was suggested by Carlo Traverso. Prasoon Tiwari was very generous in clarifying his and Ben-Or's results. Two anonymous referees provided many valuable comments and suggestions. Their efforts have significantly improved the presentation.

This report describes research done jointly at the Artificial Intelligence Laboratory and the Laboratory for Computer Science of the Massachusetts Institute of Technology. Support for the Laboratory's artificial intelligence research is provided by the Advanced Research Projects Agency of the Department of Defense under Office of Naval Research contract N00014–86–K–0180. Work done in the Laboratory for Computer Science was supported by the Advanced Research Projects Agency of the Department of Defense under Office of Naval Research contract N00014–83–K–0125.

Preparation of this report was supported by the Advanced Research Projects Agency of the Department of Defense under Office of Naval Research contract N00014–86–K–0591, the Mathematical Sciences Institute under contract U03–8300, the National Science Foundation through contract DMC–86–17355 and the Office of Naval Research through contract N00014–86–K-0281.

References

Ben Or, M. and Tiwari, P. (1988). A Deterministic Algorithm for Sparse Multivariate Polynomial Interpolation. *20th Symposium on the Theory of Computing*, 301–309.

Brown, W. S. (1971). On Euclid's Algorithm and the Computation of Polynomial Greatest Common Divisors. *J. Assoc. Comp. Mach.* **18/4**, 478–504.

Gantmacher, F. R. (1959). *The Theory of Matrices*, Chelsea Publ., New York.

von zur Gathen, J. (1983). Factoring Sparse Multivariate Polynomials. *Proceedings, IEEE Symposium on the Foundations of Computer Science*, 172–179.

von zur Gathen, J. (1985). Irreducibility of Multivariate Polynomials. *J. of Computer and System Sciences* **31**, 225–264.

von zur Gathen, J., Kaltofen, E. (1985). Factoring Sparse Multivariate Polynomials. *J. of Computer and System Science* **31**, 265–287.

Grigoriev, D. Yu. and Karpinski, M. (1987). The Matching Problem for Bipartite Graphs with Polynomially Bounded Permanents is in NC. *Proceedings of the 28th IEEE Symposium on the Foundations of Computer Science*, 166–172.

Heintz, J., Sieveking, M. (1981). Absolute Primality of Polynomials is Decidable in Random Polynomial Time in the Number of Variables. *Lecture Notes in Computer Science, vol. 115*, 16–28.

Kaltofen, E. (1985a). Polynomial-Time Reductions from Multivariate to Bi- and Univariate Integral Polynomial Factorizations. *SIAM J. of Computing* **14**, 469–489.

Kaltofen, E. (1985b). Computing with Polynomials given by Straight-Line Programs I; Greatest common divisors. *Proceedings, 17th ACM Symposium on Theory of Computation*, 131–142.

Kaltofen, E. (1987). Factorization of Polynomials Given by Straight Line Programs..

Kaltofen, E., Yagati, L. (1988). *Improved Sparse Multivariate Polynomial Interpolation Algorithms*, Report 88–17, Department of Computer Science, Rensselaer Polytechnic Institute, Troy, NY.

Moses, J., Yun, D. Y. Y. (1973). The EZGCD Algorithm. *Proceedings of ACM National Conference*, 159–166.

Pólya, G., Szegö, G. (1976). *Problems and Theorems in Analysis*, Springer-Verlag, New York.

Press, W., Flannery, B. P., Teukolsky, S. A., Vetterling, W. T. (1986). *Numerical Recipes, The Art of Scientific Computing*, Cambridge University Press, Cambridge.

Rosser, J. B., Schoenfeld, L. (1962). Approximate Formulas for Some Functions of Prime Numbers. *Illinois Journal of Mathematics* **6**, 64–94.

Schwartz, J. T. (1980). Probabilistic Algorithms for Verification of Polynomial Identities. *J. Assoc. Comp. Mach.* **27**, 701–717.

Tiwari, P. (1987). *Parallel Algorithms for Instance of the Linear Matroid Parity with a Small Number of Solutions*, IBM Research Report RC 12766, IBM T. J. Watson Research Center, Yorktown Heights, NY.

van der Waerden, B. L. (1953). *Modern Algebra*, F. Ungar Publ. Co., New York.

Wang, P. S.–H., and Rothschild, L. P. (1975). Factoring Multivariate Polynomials over the Integers. *Math. Comp.* **29**, 935–950.

Wang, P. S.–H. (1978). An Improved Multivariate Polynomial Factoring Algorithm. *Math. Comp.* **32**, 1215–1231.

Zippel, R. E. (1979). Probabilistic Algorithms for Sparse Polynomials. *Lecture notes in Computer Science 72: Symbolic and Algebraic Computation*, 216-226.

Zippel, R. E. (1980). Newton's Iteration and the Sparse Hensel Algorithm. *Proceedings of SYMSAC'80*, 68–72.